AF710944

. . . *with appreciation and compliments of the*

American Society of Agronomy

Crop Science Society of America

Soil Science Society of America

Sept 13, 1989

Intellectual Property Rights Associated with Plants

Intellectual Property Rights Associated with Plants

Section I. Policy Issues Relating to Proprietary Rights

Editors
B. E. Caldwell and J. A. Schillinger

Section II. An Evaluation of the Issues, Challenges, and Opportunities Related to Plant Patenting

Editors
J. H. Barton, C. O. Qualset, D. N. Duvick, and R. F Barnes

Steering Committee
C. O. Qualset and D. N. Duvick, cochairs
R. F Barnes, J. H. Barton, R. L. Lower, R. K. Perrin, L. J. Raines, H. L. Shands, H. Silverstein, and S. B. Williams, Jr.
P. J. Fitzgerald, ex officio

Editor-In-Chief CSSA	*Editor-In-Chief ASA*	*Editor-In-Chief SSSA*
C. W. Stuber	G. H. Heichel	D. E. Kissel

Managing Editor	*Associate Editor-at-Large*
S. H. Mickelson	Annette J. Tallard

ASA Special Publication Number 52

Crop Science Society of America, Inc.
American Society of Agronomy, Inc.
Soil Science Society of America, Inc.
Madison, Wisconsin, USA.
1989

Cover Design: Patricia Jeffson

Crop Science Society of America, Inc.
American Society of Agronomy, Inc.
Soil Science Society of America, Inc.
677 South Segoe Road, Madison, WI 53711, USA

Copyright © 1989 by the Crop Science Society of America, Inc.
American Society of Agronomy, Inc.
Soil Science Society of America, Inc.

ALL RIGHTS RESERVED UNDER THE U.S. COPYRIGHT LAW OF 1978 (P.L. 94-553)

Any and all uses beyond the limitations of the "fair use" provision of the law require written permission from the publisher(s) and/or the author(s); not applicable to contributions prepared by officers or employees of the U.S. Government as part of their official duties.

Library of Congress Cataloging-in-Publication Data

Intellectual property rights associated with plants.

(ASA special publication; no. 52)
Comprised in part of papers presented at a workshop held Jan. 31–Feb. 3, 1989, in Anaheim, Calif.
1. Plants, Cultivated—United States—Varieties—Patents—Congresses. I. Caldwell, B. E. (Billy E.) II. Crop Science Society of America. III. Series.
S1.A453 no. 52 630 s [346.730486] 89-22397
[KF3133.P53] 630 s [347.306486]
ISBN 0-89118-101-6

Printed in the United States of America.

CONTENTS

Part B. Plant Patenting Workshop Report

FOREWORD

Plant varieties (cultivars, see Editor's note) have been patentable in the USA since the Plant Patent Act of 1930, which allowed patenting of asexually propagated varieties. Since 1970, sexually propagated varieties could be protected under the Plant Variety Protection Act. Processes in plant breeding have been protected under utility patents for many years, but it was not until the modern era of biotechnology that organisms—including microorganisms, plants, and animals—could be protected by utility patents. Landmark decisions include the U.S. Supreme Court case (1980) *Diamond v. Chakrabarty* and the U.S. Patent and Trademark Office (1985) action *Ex Parte Hibberd*, which permitted patents on organisms, traits, and genes.

Understandably, these recent developments and the explosion in technology to manipulate genes by molecular means have raised a number of questions about the interpretation and application of the U.S. Code dealing with patents, including the Plant Variety Protection Act of 1970. The environment for research and development (R & D) in crop improvement is changing rapidly, both in public and private sectors. In the public sector, plant breeding continues but with a dramatic shift in staffing to reflect development and adoption of new technologies—molecular plant breeding is a near-reality. At the same time, the dynamics of private sector R & D encourage infusion of venture capital into biotechnology and various degrees of concentration and vertical integration in the seed development and marketing industry. Public-private relationships also are evolving, from the traditionally free distribution of public germplasm to the private sector, to more restricted releases, including licensing agreements. Public institutions are exploring opportunities to recover costs of R & D and to generate income for student training through patenting of processes, plant genes and gene systems, and varieties.

All of these factors, and more, have caused a great deal of uncertainty and uneasiness among scientists, research managers, public and private variety development organizations, farmers, and consumers.

The Crop Science Society of America (CSSA) provided a forum for discussion and debate of plant breeders' rights in 1963 [A.A. Hanson and M.G. Weiss (ed.) 1964. Plant breeders' rights. ASA Spec. Publ. 3. ASA, CSSA, Madison, WI] and variety protection in 1968 [R.V. Frakes (program chair). 1969. Variety protection by plant patents and other means. CSSA Technical Series 1. CSSA, Madison, WI] as these topics became important to developers of crop varieties. The CSSA continued this dialogue at its annual meeting in 1988, results of which are published in the first section of this special publication. During that symposium, it became clear that additional review and discussion was needed to clarify existing policy and practice of proprietary protection as it exists in the USA and how it may impact upon public and private plant breeding, international trade, the sharing of genetic resources, and the public good.

The CSSA was joined by its sister societies, the American Society of Agronomy and the Soil Science Society of America as well as the American Agricultural Economics Association and the American Society for Horticultural Science in sponsoring an evaluation workshop 31 Jan. to 3 Feb. 1989. The workshop was supported in part by a grant from the USDA. The second section of this special publication is devoted to invited papers presented during the workshop on the biological bases for crop variety development and the legal bases for the various options for proprietary protection of plants and the Workshop Report.

We are pleased with the great interest in the whole field of intellectual property rights, and hope that this publication will stimulate further developments in legal and practical aspects of proprietary protection of plants. Especially needed is an expanded educational effort. To that end we are confident that the papers included in this special publication will be widely used.

We thank the authors and editors for their efforts in producing this publication.

Calvin O. Qualset, *president*
Crop Science Society of America

Edward C. A. Runge, *president*
American Society of Agronomy

John J. Mortvedt, *president*
Soil Science Society of America

Editor's note: The terms variety and cultivar are considered to be synonymous; which term used depends on an authors' preference, and in some instances, on legal requirements.

CONTRIBUTORS

Stephen A. Bent	Partner, Foley & Lardner, Schwartz, Jeffery, Schwaab, Mack, Blumenthal & Evans, Alexandria, VA 22314
Fredrick A. Bliss	Department of Pomology, University of California, Davis, CA 95616
Kenneth H. Evans	Plant Variety Protection Commissioner, Plant Variety Protection Office, USDA, Beltsville, MD 20705
Jeffrey L. Ihnen	Robbins & Laramie, Washington, DC 20037
Robert J. Jondle	Bioscience Advisor, Robbins & Laramie, Washington, DC 20037
Roger A. Kleese	Formerly, Vice President, Plant Research and Development, Molecular Genetics, Inc. Currently, President, Plant Science Research, Inc., Minnetonka, MN 55343
A. David Kline	Professor and Chair, Department of Philosophy, Iowa State University, Ames, IA 50011
Victor L. Lechtenberg	Associate Director, Agricultural Experiment Station, Purdue University, West Lafayette, IN 47907
William T. Schapaugh	Executive Vice President, American Seed Trade Association, Washington, DC 20005
William A. Stiles, Jr.	Staff Director, House Agriculture Committee, Subcommittee on Department Operations, Research, and Foreign Agriculture, Washington, DC 20515
William H. Tallent	Assistant Administrator, USDA-ARS, Washington, DC 20250
Eldon E. Taylor	Plant Variety Senior Examiner, Plant Variety Protection Office, USDA, Beltsville, MD 20705
Charles F. Warren	Supervisory Patent Examiner, Examining Group 180, Biotechnology, U.S. Patent and Trademark Office, Washington, DC 20231
Kenneth A. Weber	Townsend and Townsend, San Francisco, CA 94105
Sidney B. Williams, Jr.	Associate Patent Counsel, The UpJohn Company, Patents and Trademarks, Kalamazoo, MI 49007

SECTION I

Policy Issues Relating to Proprietary Rights

PREFACE

With the emergence of genetic engineering, dramatic changes have and will continue to occur in the area of plant breeding. The ability to manipulate plant genetics through genetic engineering provides new opportunities for protecting these innovations.

Private sector interest and investment have increased dramatically in crop variety development, especially for self-pollinated species. In the public sector, some plant breeding and genetic programs were redirected at the expense of traditional programs. With the increased use of patents, the role of seed certification has been questioned. Also, the ability to obtain plant patents provides an opportunity for conflict with the Plant Variety Protection Act. In addition, consideration must be given to the acceptance and utilization of patented materials by growers. The ethical questions of who benefits from proprietary rights and at what cost are critical.

The symposium entitled "Policy Issues Relating to Proprietary Rights" at the 1988 meetings of the Crop Science Society of America was designed to deal with the many issues surrounding proprietary rights. The scope of the symposium was to focus on the technical, policy, and ethical issues related to proprietary rights. The program gave members of the Society an opportunity to understand the technical aspects of the various mechanisms of protecting plant genetic innovations. The scope of the Plant Variety Protection Act and the various aspects of patenting were explained. Also, the symposium provided the opportunity for the expression of a broad spectrum of views regarding proprietary rights. It is important that all components of the plant breeding industry understand the role of each component. The symposium also provided an opportunity for participants to become aware of Congressional views as well as the views of those individuals interested in the ethics of genetic engineering.

Finally, the symposium provided insights to the participants from the experiences of individuals in a university community as well as a private company. This section contains the papers presented by the symposium participants and is intended to serve as a reference source for those who were not present, as well as to provide information for future discussions by professionals and administrators.

We want to thank the symposium speakers for agreeing to prepare their comments for publication. Their contributions made the symposium a success.

Editors

Billy E. Caldwell,
*North Carolina State University,
Raleigh, North Carolina*

John A. Schillinger,
*Asgrow Seed Company,
Kalamazoo, Michigan*

1 Overview and Status of Plant Proprietary Rights

Robert J. Jondle
Robbins & Laramie
Washington, DC

Each year, companies invest millions of dollars in biotechnology and plant breeding research with the goal of producing and marketing new biotechnology methods, genes, plant cultivars, and hybrids. Effective protection of these new plant products is necessary to provide the incentive for this large research investment. Also, an increasing number of university administrators and professors are considering using plant proprietary rights as one way to increase their research resources.

BRIEF HISTORY OF PLANT PROPRIETARY RIGHTS

Prior to 1930, no intellectual property rights were available for plants. For most plant species, plant breeding research and cultivar development were conducted by state and federal government experiment stations. Financial incentives were not present to encourage private companies to invest substantial amounts in research for developing new cultivars. In 1930, the Plant Patent Act (PPA)[1] was enacted into law. The plant patent grants the breeder the exclusive right, for 17 yr, to propagate the patented plant by asexual reproduction. The purpose of the PPA was to encourage research investment in asexually reproduced plant species.

Since 1930, over 6000 plant patents have been issued by the Patent and Trademark Office, primarily for fruit trees, flowers, ornamental trees, grape, and other horticultural species. The PPA has successfully encouraged the development of new genotypes by private industry. Since 1930, the primary development in many species of asexually reproduced cultivars has shifted from state and federal government experiment stations to private companies.

The Plant Variety Protection Act (PVPA)[2] was enacted in 1970 to provide patent-like protection for seed plant species. Prior to 1970 plant breeders in industry worked primarily with corn (*Zea mays* L.) and sorghum

[1] Townsend-Parnell Plant Patent Act of 23 May 1930 (Public, no. 245, 71st Congress).
[2] Plant Variety Protection Act of 24 Dec. 1970 (84 Stat. 1542) (7 U.S.C. 2321 et seq.).

Copyright © 1989 Crop Science Society of America, American Society of Agronomy, Soil Science Society of America, 677 S. Segoe Rd., Madison, WI 53711, USA. *Intellectual Property Rights Associated with Plants*, ASA Special Publication no. 52.

[*Sorghum bicolor* (L.) Moench], with some efforts directed to alfalfa (*Medicago sativa* L.), cotton (*Gossypium hirsutum* L.), sugarbeet (*Beta vulgaris* L.), and certain vegetables. There was little incentive to breed nonhybrid crops, or for foundation seed companies to develop and sell new inbreds, because after the sale of seed anyone could increase and sell seed of the new cultivar or line. The purpose of the PVPA was to stimulate private plant breeding research and to provide better seed cultivars to farmers and gardeners. The PVPA has accomplished these goals for certain plant species. A monograph entitled *The Impacts of Patent Protection on the U.S. Seed Industry and Public Plant Breeding* (Butler & Marion, 1985) concluded that PVPA had stimulated the development of new cultivars of soybean [*Glycine max* (L.) Merr.] and wheat (*Triticum aestivum* L.).

Utility patents have been issued on methods and processes involving plants for 30 yr, and more recently for plant cultivars and hybrids per se for over 3 yr. In 1980, the Supreme Court held that microorganisms were patentable subject matter. In 1985, the Board of Patent Appeals and Interferences (BPAI) in *Ex Parte Hibberd*[3] decided that plants could be protected with utility patents. It is expected that utility patent protection will stimulate additional research efforts toward new inventions in biotechnology and plant germplasm development.

U.S. PROTECTION OPTIONS

There are five main categories of plant protection in the USA: (i) trade secrets, (ii) contracts, (iii) PPA, (iv) PVPA, and (v) utility patents.

Trade Secrets

Trade secrets are governed by state law, and the strength of trade secret protection can vary from state to state. A trade secret can include any information, process, or germplasm that gives a company a competitive advantage. The trade secret must be held secret and confidential, with no public disclosure. Conducting tours of research facilities and laboratories may result in the public disclosure of a trade secret. In the seed industry, accidental disclosure can be a serious problem. Under trade secret law, if a company sells a bag of hybrid seed that contains a few seeds of the inbred parent, this constitutes public disclosure and the trade secret protection may be lost. For most crop species, trade secret protection may be appropriate for proprietary laboratory procedures and plant breeding research procedures. If a second party independently develops the trade secret and obtains utility patent protection, the second party can prevent the original trade secret owner from using the trade secret.

[3] 227 USPQ 443 (BPAI, 1985).

Contracts

Contracts provide a second type of plant protection. Examples of contracts are licensing agreements, secrecy agreements, condition of sale agreements, and "restricted use" labels on commercial seed bags. Contract law, like trade secret law, is governed by each state, and therefore can vary from state to state.

A restricted use label on a hybrid seed bag generally specifies that any parental seed lines found in the bag may not be used in breeding research or seed production. The level of protection that a restricted use label provides is uncertain. Under contract law, two issues weaken this protection: (i) no signature by the buyer, and (ii) the buyer has no option other than to purchase the seed with the label, therefore there is no negotiation between the buyer and seller. The strength of protection obtained by using a label will have to be defined by court challenges. License and condition of sale agreements will provide more protection than restricted use labels on seed bags. The contract protects against abuse by only the signer of the contract. If the seed or process is obtained by a third party, the only recourse is against the signer of the contract.

Plant Patent Act

The third type of plant protection is under the PPA of 1930, which is available only for asexually reproduced plants. Under the PPA, breeders who develop new and distinct asexually propagated plants can prevent others from selling plants that are asexually propagated from the patented plant for a period of 17 yr. Since 1930, over 6000 plant patents have been issued by the Patent and Trademark Office. Plant patents have been issued, for example, on flowering plants, ornamental and fruit trees, nut trees, grapes, and certain vegetable crops. Plant patents cannot be obtained for seeds, tubers, plant parts, biotechnology processes, recombinant DNA, or genes.

Plant Variety Protection Act

The PVPA was passed by Congress in 1970 to encourage the development of new, sexually reproduced cutlivars and inbred parents of hybrids. The breeder is granted a PVP certificate, which provides protection for 18 yr for a specific genotype. Since 1970, over 2100 PVP certificates have been issued. The PVP certificates are generally easier to obtain than utility patents. Currently, any trait difference(s) between two cultivars will allow both cultivars to receive PVPA protection.

The PVP certificates have both a crop exemption (usually called a farmer's exemption) and a research exemption. A farmer qualifies for this exemption if his/her "primary farming occupation is the growing of crops for sale for other than reproductive purposes." The word *primary* can be interpreted to mean that if 51% of his/her total crop is not sold as seed for reproductive purposes, the farmer qualifies for this exemption. The PVPA farmer's exemption has resulted in a series of lawsuits involving farmers with substantial seed sales of cultivars having PVPA protection. The purpose of the farmer's exemption was to allow a farmer the right to save seed and either plant it himself/herself the next year or sell it to other farmers. The American Seed Trade Association (ASTA) has recommended narrowing the interpretation of the PVPA farmer's exemption to prevent abuses by farmers. Congressional approval is required to make changes to the PVPA regarding the farmer's exemption (B.E. Caldwell, 1989, personal communication).

The PVPA also has a research exemption that allows anyone to use PVPA lines in the laboratory or in a breeding program to develop new lines. For cultivars such as soybean and wheat, the PVPA-protected line is commercially sold and available for developing new lines. One major restriction to the research exemption results from the Plant Variety Protection Office's policy of not allowing access to its seed deposits. Seed of a PVPA-protected inbred parent of a hybrid is not accessible from the Plant Variety Protection Office and is not generally available from the developing company, since it is proprietary. The hybrid of cross-pollinated crops using the PVPA parental line(s) is commercially sold but the hybrid itself cannot receive PVPA protection. Seed submitted with PVPA applications are retained by PVPA as a voucher seed sample for the duration of the certificate and then turned over to the National Seed Storage Laboratory to become a part of its general collection (Commissioner Ken Evans, 1989, personal communication). Thus, under current Plant Variety Protection Office policy, the protected PVPA line (hybrid parent) is not directly available for research programs until after the 18-yr duration of the PVP certificate.

Utility Patents

Utility patents have been issued on plant methods and processes for over 30 yr. Utility patents have been issued on plant cultivars and hybrids per se since 1985. To receive a plant utility patent, the invention must be new, useful, and nonobvious. The patent specification must contain a full disclosure of the invention, which enables anyone skilled in the art to produce the invention. For plant cultivars, the full disclosure requirement can be satisfied by making a seed deposit, which must be accessible by the public.

Table 1-1. Examples of issued utility patents.

Sunflower (*Helianthus annuus* L.) regeneration
Potato (*Solanum tuberosum* L.) cultivar
Inbred corn line
Hybrid corn line
Tryptophan overproducer mutants
Mushroom (*Agaricus bisporus* L.) mutant strains
Process producing odorless soybean
Bean (*Phaseolus vulgaris* L.) plant having low pod detachment force
Squash (*Cucurbita pepo* L.) cultivar
Tetraploid corn and method to produce
Process for microplant propagation
Herbicide resistance in tobacco (*Nicotiana tabacum* L.)
Kiwi (*Actinidia chinensis* Planch.) fruit plant

Plant utility patent protection lasts for 17 yr after issuance, and is available for biotechnology processes, genes, seeds, plant parts, cultivars, and hybrids. Examples of recent plant utility patents that have been issued are shown in Table 1–1. Utility patents have no farmer's exemption and no research exemption for the development of new cultivars. During the 17-yr protection period, the seed deposit is accessible by the public and can be used for testing purposes but not in breeding programs to develop new commercial lines. After the expiration of the 17-yr protection period, the seed or process is in the public domain and can be used to develop new lines.

A COMPARISON BETWEEN PLANT VARIETY PROTECTION ACT AND UTILITY PATENTS

The PVPA and utility patent protection are compared in Table 1–2. Both PVPA and utility patents require the invention to be new or novel. The PVPA requires the cultivar or inbred to be distinctive and uniform. Currently, PVPA distinctiveness is satisfied if any trait, regardless of the trait's importance, differs from all previous lines. The issue of whether PVPA should have a well-defined minimum distance requirement between protected lines is being reviewed by committees of the American Seed Trade Association and proposals are being developed to present to the Plant Variety Protection Office. Congressional approval on changes to the PVPA will probably be required to allow the Plant Variety Protection Office to define and use ASTA-proposed minimum distances criteria as a basis to reject a PVPA application.

For utility patents, close look-alike hybrids and cultivars generally may not both obtain patent protection under the doctrine of equivalents. The doctrine of equivalents states that if the product accomplishes the same things as the patent claims and in substantially the same way, there is infringement. In this situation, the first patented hybrid may bar patent protection on the look-alike hybrid or cultivar. Patent laws require the invention to be useful and nonobvious to one of ordinary skill in the art. Predictability is an important aspect of nonobviousness. If an invention is predictable to one skilled in the art, it will not be patentable.

The scope of protection for PVPA and utility patents is compared in Table 1–2. Under PVPA law, the seed and plant of an inbred or cultivar are protected. With patent protection, seeds, whole plants, plant parts, genes or physical traits, cultivars, hybrids, methods of plant regeneration, and other biotechnology processes and products are all patentable inventions. The PVPA protection is not afforded hybrids, fungi, or bacteria, whereas patents can protect plants, microbes, and animal inventions. The PVPA protection lasts a total of 18 yr, compared to 17-yr protection with utility patents.

The PVPA seed deposit is not accessible by the public according to current Plant Variety Protection Office policy. After the 18-yr duration of the certificate, the seed becomes part of the National Seed Storage Laboratory general collection. Utility patent seed deposits must be accessible by the public as soon as the patent issues. During the life of the patent, the public can

Table 1-2. A brief comparison of U.S. plant variety protection certificates and utility patents for plants.

	Plant variety protection certificates	Utility patents for plants
Requirements	Novelty, uniformity, distinctiveness *Novelty:* new, has not previously been known publicly *Uniformity:* any variations are describable, predictable *Stability:* remains unchanged when sexually reproduced *Distinctiveness:* differs by one or more characteristics	Novelty, utility, nonobviousness *Novelty:* new, has not previously been known publicly *Utility:* usefulness for other than research purposes *Nonobviousness:* not an obvious variation of the known art
Scope of available protection	Seed of described inbred or cultivar, "any novel cultivar of sexually reproduced plant, other than fungi, bacteria, or first generation hybrids"	Seed, plants, plant parts, genes, or physical traits for inbreds, hybrids, and cultivars; methods of developing inbreds, hybrids, and cultivars; methods of plant regeneration
No. years protection	18 yr from issue date of certificate	17 yr from issue date of patent
Right to	Exclude others from selling, importing, exporting, reproducing, etc., for term of certificate plus time period from distribution of cultivars with proper notice to issue date	Exclude others from making, using, or selling for term of patent
Seed deposit	Seed deposit is not accessible by others according to present Plant Variety Protection Office policy	Seed deposit is accessible, but patent owner can be notified of who received seed. Seed can be used for research only and not to develop new commercial products without permission (license) from patentee
Compulsory licensing	None, except if Secretary of Agriculture determines to be in public interest	None
Cost	Government fee of $2400 per application + cost of preparation and amendment (if any)	Government fee of $500 to 1000 + cost of preparation and amendment (if any) + maintenance fees at 3½, 7½, and 11½ yr from issue date
Business restrictions	Can use protected inbred/cultivar to develop new inbreds, cultivars, and hybrids; farmer's exemption	Will protect multiple inbreds, hybrids, or cultivars having patented characteristic so newly developed inbred/cultivar will be covered; seed can be used for basic research but not to develop commercial products without permission (license) from patentee; no farmer's exemption

use the seed only for testing purposes and not to develop new commercial products, unless they have permission (license) from the patent owner. When the 17-yr patent protection period expires, the seed must be freely available to the public to use in research or to develop new cultivars.

For biotechnology processes, recombinant DNA, and genes, the patent specification must be complete enough to enable one skilled in the art to develop the patented invention. To receive the benefit of patent protection, the inventor must fully disclose the invention to the public.

The last major difference between the PVPA and patents is that the PVPA has both farmer's and research exemptions, while patents have neither of these exemptions.

INTERNATIONAL PLANT PROPRIETARY RIGHTS

In 1961, the Paris Convention established the UPOV (International Convention for the Protection of New Plant Varieties) to establish the breeder's rights for plant cultivars. Currently, 21 countries have modeled their statutes after UPOV, including most of the European community, Argentina, Japan, and the USA. Both sexually reproduced and vegetatively propagated plants can be protected. A separate application must be filed to receive protection in each member country. The UPOV Convention requires that each cultivar have a specific, unique name that is used when selling the cultivar. In most European countries, only certain plant species permit use of the protected cultivar in research programs to develop new cultivars. In addition, some countries have mandatory licensing of the variety to any individual or company upon payment of a reasonable royalty fee.

Utility patent protection for plant cultivars and hybrids is not available in Argentina, Brazil, Canada, China, India, the Philippines, and the USSR. The current policy of the European Patent Office (EPO) is that plant cultivars and hybrids per se are not patentable. The EPO will grant utility patents for the biotechnology processes of inserting a desired gene or for the gene itself that can be inserted in many different plant cultivars, or for a plant having a new trait. In a recent report, a European commission has recommended that patent protection be granted to plants unless the plant is a fixed cultivar. The patent would then be dominant over the breeder's rights. In the USA, plant cultivars, inbreds, and hybrids can be protected by utility patents. Australia, France, and West Germany provide utility patent protection for plant species not covered by the PVPA.

ISSUES INVOLVING PLANT PROPRIETARY RIGHTS

Issues affecting plant proprietary rights continue to change and evolve, both in the USA and internationally. In this section, some of the more commonly discussed issues are addressed. For each issue, the perspective can vary substantially depending on the crop species involved and the organization's goals. Federal agencies, public universities, private universities, large or small seed and nursery companies, and biotechnology companies—each can have differing viewpoints on many of these issues. To further complicate this discussion, each crop species often results in a different perspective depending upon whether the species involved is a field crop, horticultural, or forestry species.

Effect on Third World Countries

The effect of PVPA, UPOV breeder's rights, and utility patents on Third World countries is expected to be minimal. Plant proprietary rights and plant protection are not recognized in most developing countries. Seed of superior plant cultivars and hybrids tends to become distributed in these developing countries, where there is no protection, and the seed can be used freely for production and research purposes. Because private companies are generally reluctant to sell seed in countries with no proprietary rights, the seed may reach the developing country somewhat later. Once the seed becomes available to the farmers and producers in Third World countries, there is no royalty or fee to pay. For seed companies, this highlights the difficulty in protecting proprietary germplasm worldwide, especially when cultivars are obtained and modified in countries that do not have any type of plant protection.

Obtaining Both Plant Variety Protection Act and Utility Patent Protection

A breeder currently can receive both a PVPA certificate and a utility patent on a plant cultivar or inbred. While this combination would logically appear to provide stronger protection than either type alone, holding both a PVP certificate and a utility patent may actually weaken the protection. There is a risk the courts may view such acts as *double patenting* which may reduce the level of protection for the cultivar. This matter will have to be decided eventually by court challenges. A double patenting problem also exists when receiving both PPA and utility patent protection for the same genotype.

Advantages and Disadvantages of Each Protection Type

The PPA protection is available for only asexually reproduced plants and not for seeds, plant parts, or biotechnical products and processes. A PPA is relatively simple to obtain from the Patent Office. The inventor must submit a specification listing the characteristics that distinguish the invention and point out where and in what manner the plant cultivar has been asexually reproduced. The protection covers the cultivar only, and new cultivars with minor differences can also receive plant patent protection. The PPA protection is quite limited, but is easier to obtain than a utility patent.

The PVP certificate has the advantage of protecting seeds of cultivars and inbreds. This protection is not allowed under PPA. Like the plant patents for asexually reproduced plants, the PVP certificate is relatively easy to obtain. The applicant must file an application listing specified traits and a seed deposit of 2500 seeds. The PVPA protection is weakened by (i) a research exemption that allows anyone to use the PVPA-protected material in breeding programs; (ii) a farmer's exemption that currently may allow large-scale farmer-dealers to sell substantial quantities of seed of the protected line; or (iii) the lack of a distinctiveness criteria (minimum distance between genotypes), i.e., a second cultivar with a minor change in any trait would also be allowed to receive a PVP certificate. In situations where the cultivar does not merit the extra effort, or does not qualify for a utility patent, a PVP certificate provides protection for the cultivar.

Utility patent protection is the only form of plant protection available for hybrids, biotechnology, and plant breeding methods and processes, genes, recombinant DNA, and plant parts such as pollen and tissues. Utility patent protection is more difficult and expensive to obtain. One patent can have one or more claims, each protecting a separate aspect of the invention. These claims can provide a broader scope of protection than plant patents (PPA) or PVP certificates.

It is important to have all three types of plant protection options available to the inventor/developer. Which option is chosen, if any, will vary depending on the new material or process, organizational goals, and the plant species involved.

Effect on the Exchange of Germplasm

Neither the PPA nor the PVPA has restricted the exchange of germplasm; both are generally recognized as having been beneficial to U.S. agriculture. While the effect of utility patents is not fully known at this time, no major restriction on the exchange of germplasm is expected. Federal organizations, including the USDA and the ARS, will continue to allow access to newly developed germplasm. Public universities will also allow access to germplasm, possibly in certain cases charging a nominal fee. Many public institutions have the policy of not protecting their new germplasm or techniques, although there are some universities receiving protection with the intent to license at a reasonable fee and on a nonexclusive basis. Some private universities will probably seek plant protection for new processes and cultivars, but there will be a strong incentive to license on a nonexclusive basis to maximize their profit. Small- to medium-sized biotechnology companies and foundation seed companies will also generally need to license their germplasm to many organizations in order to maximize their profit. Large seed companies probably have the most incentive to exclusively control the germplasm and techniques they develop. But as more organizations have protected plant materials, all companies, regardless of size, will likely cross-license to obtain germplasm or biotechnology products from other companies. This may encourage the exchange of proprietary materials. For many crop species, including oat (*Avena sativa* L.), barley (*Hordeum vulgare* L.), etc., the profit potential often is inadequate to justify the expense of plant protection. Therefore, germplasm of most species will continue to be freely accessible. For species such as corn, soybean, sorghum, and certain vegetables and flowers, there will be an incentive to receive plant protection.

Selecting the Type of Plant Protection

The type of protection needed will vary, depending upon (i) the goals of the organization, and (ii) which plant species is involved. Obviously, an organization's goals will vary significantly depending on whether it is for public or private sectors. Federal agencies and public universities have traditionally allowed relatively free access to plant materials and information developed by their research staffs. These organizations will generally decide to use none of the plant protection options for most species. In certain cases, public organizations may choose to protect a few species or techniques and market these in a nonexclusive manner for a reasonable fee. Such protection allows a modest return to cover the expenses of the protection and handling costs involved, and possibly have some funds for future research projects. Private companies working with high-profit crops and spending large amounts in plant research will have more incentive to protect their new inventions and germplasm. Again, many plant species do not have an adequate profit potential to support obtaining plant protection.

Determining the Scope of Plant Protection

Under the PPA and PVPA, only the exact genotype described receives protection. For utility patents, the individual claims of the patent define the invention and identify the protection given the patent holder. Therefore, plant protection with utility patents may be broader in scope than with PPA and PVPA protection.

Defining the Doctrine of Equivalents

To determine if a patent is infringed, both the claims of the patent and the doctrine of equivalents must be considered. The claims specifically define what is protected by the patent. If a second product is similar to, but does not fall within the bounds of the patent claims, it may still infringe the patent. The doctrine of equivalents states that if the product accomplishes the same thing as the patent claims and in substantially the same way, then there is infringement. For example, with plant cultivars per se, the claims will be narrowly defined to cover a cultivar with traits identified in the patent. Using the doctrine of equivalents, certain close look-alike cultivars may infringe the patent. If challenged, the final determination would need to be defined by the court. For plant patents covering new processes, the doctrine of equivalents will generally apply more broadly than with individual cultivars and hybrids per se. Currently, neither the PPA nor the PVPA provide any protection broader than the exact genotype described. Any trait modification will allow the developer to sell the modified genotype.

Defining Nonobviousness

Nonobviousness is determined by (i) comparing the proposed invention with what is currently known in publications and other prior art, (ii) assessing the level of skill of the average person skilled in the art, and (iii) determining if the invention would be obvious to one of average skill in the art. Predictability is an important aspect of nonobviousness. To the extent that the new invention is predictable, it will not be patentable. Nonobviousness is initially determined by the Patent Office examiners when reviewing the patent application, but any contested Patent Office decisions or patents are finally decided by the courts.

CONCLUSION

There are various options for plant protection, each with its advantages and disadvantages. The important goal is to have adequate protection, which may require the use of one or more of these options. Many fears or concerns have arisen because of the lack of adequate knowledge of patent law. It is anticipated that utility patents, as well as other protection options, will not adversely impact on the seed industry but will enhance further innovations.

REFERENCE

Butler, L.J., and B.W. Marion. 1985. The impacts of patent protection on the U.S. seed industry and public plant breeding. North Central Regional Res. Publ. 304. North Central Project 117. Monograph 16. Res. Division, College of Agric. and Life Sciences, Univ. of Wisconsin, Madison, WI.

2 The Seed Trade's View on Proprietary Rights

William T. Schapaugh

American Seed Trade Association
Washington, DC

I have been asked to present the seed industry's view on proprietary rights. My comments will be limited to protection under the plant variety protection (PVP) system and utility patent protection for the seeds/plant genetics supply industry.

About a year ago, the Board of Directors of the American Seed Trade Association (ASTA) directed that a survey of members be made for the purpose of developing key concepts so that ASTA could have a voice in the decision-making on the nature and scope of intellectual property protection for our industry.

A Discussion Paper and Questionnaire was developed and mailed to members. After thorough consideration of responses from members and a great deal of debate, the ASTA Executive Committee approved in November 1987, by majority vote, a Statement of Basic Concepts and Principles for Patent and PVP rights for the Seed Industry. The Executive Committee also asked that another paper be developed summarizing the responses of members—both majority and minority. This second paper also addresses more fully the topic of patenting results of conventional or traditional methods of plant breeding. Copies of both papers are available from the ASTA office.

The seed industry is diverse. There is also diversity and an element of complexity inherent in the subject of patenting and PVP for our industry. And clearly, a most important consideration, the issues involved are further subject to personal opinions about what is fair, what is right, and what is in the best interests of agriculture—what is the best public policy. It is therefore not surprising that responses and opinions of members of the industry are not unanimous. In fact, opinions within a company are not necessarily the same, as, for example, those between breeders and patent attorneys.

The ASTA Statement has a worldwide orientation or purpose because (i) of the needs for expansion, changes to, and harmonization of intellectual property right laws around the world; (ii) the seed industry is international in nature; and (iii) issues related to germplasm conservation, exchange, and utilization are worldwide in scope.

Copyright © 1989 Crop Science Society of America, American Society of Agronomy, Soil Science Society of America, 677 S. Segoe Rd., Madison, WI 53711, USA. *Intellectual Property Rights Associated with Plants*, ASA Special Publication no. 52.

AMERICAN SEED TRADE ASSOCIATION MAJORITY MEMBER VIEWPOINT

For this presentation I want to highlight the key concepts and principles for patenting, and PVP rights for the seed industry as approved by our Executive Committee. I will discuss first the statement as approved, then I will comment on minority viewpoints.

The ASTA Statement, after a preamble, states the following:

Considering,

1. Due to the nature of plant breeding, continued improvements in productivity of crops is dependent on use of existing varieties and germplasm as building blocks to develop new varieties to meet the challenges and needs of the future.
2. Undue restrictions on flow and utilization of plant germplasm would undermine incentives for, and negatively impact competitive research programs to continue genetic improvements of crops.
3. Unrestricted rights under patent laws and application of patent law doctrines could result in "locking up" plant germplasm or block alternative genetic approaches to solve crop performance problems/needs.
4. The necessity and public interest of providing a framework for legal protection for inventors/breeders that (i) equitably serves as incentive for risk-taking in the form of investing in both conventional and genetic engineering research and (ii) is as free as possible from confusion, controversy, and potential litigation.
5. The highly complex and controversial scientific and legal issues involved in patenting for the seeds/plant genetic supply industry.

Therefore, ASTA supports the following fundamental concepts to be in the public interest in development and application of patent, and PVP conventions and laws on a worldwide basis.

1. New inventive genetic components of plants should be eligible to qualify for patent protection. Rights to such genetic material should extend to other plants/varieties that contain and express the trait(s) of such genetic material.
2. Processes, including essentially biological processes, should be eligible to qualify for patent protection, provided the level of human intervention in the process is decisive for causing the result. Direct products of a process should be included within the scope of a process patent. Products of a multistep process in which different products are produced from successive steps in the process should be included within the scope of a process patent only to the extent that human intervention is decisive in causing the different results of the steps involved.
3. Near-duplicate varieties differing from existing varieties in only one or a few inconsequential traits should not qualify for legal protection. Varieties per se should be protectable only under PVP laws that

are strengthened to exclude from protection cosmetically developed, plagiarized varieties; if PVP laws are not so strengthened, then varieties should be eligible to qualify for utility patent protection.

4. Farmers should have equivalent, but restricted rights to save and plant saved-seed of varieties protected by PVP and varieties that contain patented genetic components. The "right to save seed" exemption for farmers must be significantly tightened to prevent abuses that now substantially undermine breeders' rights.
5. Patents for genetic components or characteristics of crops should be granted on the following basis:
   a. Only genetic components that directly serve to cause expression of useful characteristics in crops should be eligible for protection.
   b. Characteristics of crops should not be patented unless the direct genetic causative agent is identified and itself qualifies to be patented.
   c. Alternative genetic approaches to achieve the same characteristic or trait in crops shall not infringe a prior patent.
6. In recognition of the above, and subject to no. 7 below, there should be a legally established compulsory licensing system under which a person who develops a new, distinct variety that falls within the scope of a patent of another party shall have a right to commercialize the new variety, free of infringement, under a system that provides equitable compensation to the owner of the patent.
7. Provisions for compulsory licensing would apply when the scope of a product or process patent is so broad as to encompass within its scope varieties or parts of plants not yet developed. Examples of such patents would include patents with claims directed to:
   a. Characteristics of crops,
   b. Genetic components that act as agents for expression of characteristics of crops, or that serve to regulate or control further steps of synthesis of plant material, or
   c. Processes of genetic manipulation.

Except for extraordinary instances involving national interest, individual varieties should not be required to be licensed to others.

AMERICAN SEED TRADE ASSOCIATION MINORITY VIEWPOINTS

As mentioned earlier, the seed industry is a diverse industry and differences of opinion exist. Within ASTA there are minority views on each end of the spectrum. One minority view is that no plant material, genes, or processes should be considered as patentable—in effect, that PVP protection is all that is needed.

Another minority view, held by some of the firms that invest heavily in research and that account for important sales in certain market segments, is that there should be no restrictions, conditions, or limitations on patent rights for our industry. In particular, such companies want their strongly held views to be expressed and understood. Examples of such views would include:

a. Opposition to any change in U.S. law that would limit patent rights for our industry.
b. Belief that varieties should be eligible for patent protection.
c. Belief that mandatory licensing is neither necessary nor desirable, and that mandatory licensing instead leads to lower returns on research and to less research being conducted.

RECENT DEVELOPMENTS

Next, I will discuss developments concerning intellectual property rights since the ASTA position was approved.

First, ASTA is one of a number of organizations that make up ASSINSEL, the international association that represents private plant breeding interests and whose membership includes representation from many countries around the world. The ASSINSEL met in early June in the United Kingdom and approved a paper of concepts regarding protection of biotech inventions. The ASSINSEL paper is consistent in many respects with the ASTA paper. Although there was broad support within ASSINSEL for the concepts adopted, there were minority opinions.

Secondly, permit me to make a few comments on protecting varieties and what is a variety. Some believe that protecting varieties by utility patents represents stronger protection than PVP. I am not sure that is correct, and it is not my intent to debate the point here. For example, under patent law, there is not the same kind of research exemption as in PVP. There is an "experimental use" doctrine that has developed under patent law, but application of that doctrine to plant breeding is uncertain. Some attorneys argue that the experimental use doctrine permits experimentation only to satisfy scientific curiosity. As long as there is uncertainty, patenting of varieties could lead to questions, confusion, and potential litigation about what a breeder may or may not do with a patented variety without infringing the rights of others. How much confusion would there be if half of all new varieties in the past 10 yr had been protected under patent law and the other half under PVP? What would be the impact on research and development programs, public or private?

This is an important public policy issue. Just a few years ago, in addressing the "Undertaking of Plant Genetic Resources" promoted by the antibusiness forces within the Food and Agricultural Organization of the United Nations (FAO), ASTA took the position that "all improved plant germplasm offered for sale any place in the world is available for legitimate breeding purposes. . ."

Notwithstanding the foregoing comments, the ASTA Statement points out the problem of near duplicate varieties that differ from existing varieties in only one or a few inconsequential traits. This is an old, persistent problem.

How different should one variety be from another to be considered a unique, distinct variety? The question of minimum distance between varieties has never been adequately defined.

Recognizing the concern in our industry, ASTA has appointed a Steering Committee to study minimum distance. The objective is to define or develop guidelines on how different varieties should be, one from another, to be considered distinct. Nine crop species have been selected for study, and committees for each species have been established, made up of both public and private sector representation. These committees will be meeting during the rest of this year and have a target date set for next spring for their recommendations.

In this regard, it should be noted that it has been decided to open up the UPOV Convention for modification, and minimum distance or how to define distinctness is one of the key areas under review. The acronym UPOV is for the international agreement or convention under which the PVP laws of the USA and a number of other nations have been promulgated.

A recent draft for a revised UPOV Convention incorporates a new concept to the effect that if a new variety is essentially derived from an existing variety, while the new variety may be considered distinct, it could not be marketed, free of infringing the rights of the developer of the prior variety, without payment of compensation. In other words, it has been proposed that a system of dependency between varieties be introduced into the UPOV Convention. Though not yet fully defined, this concept, including the aspect of compulsory licensing and payment of remuneration, has gained considerable support among private breeders because of the need to stop what many refer to as "plagiarism" where a popular variety is changed by someone in one or a few traits to come up with a slavish imitation of the first variety.

On this most critical point, another recent development has come from the World Intellectual Property Organization (WIPO), the International Union for the Protection of Industrial Property, headquartered in Geneva together with UPOV. The WIPO has published a set of *Suggested Solutions* concerning Industrial Property Protection of Biotechnological Inventions. These *Suggested Solutions* were developed after extensive review of the issues involved with, and soliciting the views of, both governmental representatives of nations around the world and also international nongovernmental organizations, including grower organizations, seed trade interests, etc.

One *Suggested Solution* reads, in part, as follows:

> Where the holder of a plant variety right for a variety that represents a significant technical advance over a patented invention in the relevant area wants to carry out an activity concerning the said new plant variety which is within the scope of protection of the said patent, he shall have a right to obtain a license under the said patent in order to carry out such activity, subject to the payment of a reasonable remuneration.

The International Bureau of WIPO explained its position on this suggestion; excerpts follow:

> The possibility of a dependency license would promote technological progress *in the field of plants* because it would *stimulate* research in the direction of *im-*

> *provement* of patented inventions concerning plants. The interests of the patent owner are safeguarded by the requirement of reasonable remuneration. The contribution of the holder of a plant variety right should be placed . . . on an equal footing with that of the owner of a patent. In the field of functions governed by genetic material, the number of possible solutions may be . . . small(er) . . . therefore, the practical exclusivity achieved by the patent is much more meaningful. The suggested solution strikes a *fair balance between the various interests* . . .

Another important recent development is efforts to strengthen the "right to save seed, crop exemption" section of the PVP—often erroneously called the "farmer's exception." The right to save seed exemption under PVP law as currently interpreted is absolutely unsatisfactory to the seed industry. There are so many abuses taking place under this exemption that the benefits of the Plant Variety Protection Act have been seriously eroded. This problem is certainly providing an incentive to breeders to consider protecting new varieties in the USA under patents instead of PVP, because under patent law there is no farmer's exemption. To debate what rights a farmer may or may not have to use a patented variety goes beyond my purpose today. However, I do wish to convey the thought that the crop exemption under PVP law must be changed. It is a priority for our industry.

SUMMARY COMMENTS

In closing, I would like to refer briefly to several important considerations related to the importance of germplasm exchange and utilization, alternative genetic pathways to achieve crop traits, and potential affects of unrestricted patent rights on the plant breeding/seeds/genetics supply industry. All of these have motivated many within the industry to support a system of compulsory licensing and payment of royalty fees under circumstances when patents with very broad claims are issued.

I want to emphasize that within ASTA we are not suggesting mandatory licensing of varieties. Rather, we are concerned about claims that can extend to many varieties, even to an entire species. An example of the sort of broad claim that, according to the ASTA concept, would be subject to licensing to others can perhaps best be illustrated by one or two questions. Consider the development of Phytophthora root rot (*Phytophthora megasperma* Drechs.) resistance in soybeans [*Glycine max* (L.) Merr.]; would it have been in the best interest of the seed industry or the best interest of agriculture if the person who first developed, bred, or identified Phytophthora root rot resistance would have had an absolute right for 17 yr to prevent anyone else from developing and marketing other, different varieties also resistant to Phytophthora? Is it in the best interest of the public for the owner of a patent on a crop trait such as atrazine (2-chloro-4-ethylamino-6-isopropylamino-1,3,5 triazine) resistant soybean to have the right for 17 yr to prevent others from marketing other different atrazine resistant soybean varieties, even when the source of resistance is genetically completely different? Those are examples of broad claims that are permitted under patent law. Some in our industry believe a system of licensing is in the public interest and is in the best interest of agriculture.

That is the sort of motivation that has led other seed industry groups to support concepts akin to those of ASTA and ASSINSEL. In this context, I could mention the Canadian Seed Growers, the British Plant Breeders, COMASSO (the European plant breeder's organization) and several others. Also in this regard, it is interesting to note that one organization of private plant breeders in the USA has submitted a letter of protest to the U.S. Patent Office in which major concerns were expressed about the broadness of claims contained in one particular patent application. Comments such as the following were made:

> Broad claims about (X trait) and . . . (Y specie) could limit participation by others in an entire market segment, reducing the incentive for additional research.
>
> It is also a concern that a major component of . . . quality may be tied up in a broad patent. It is conceivable that issuance of this patent could be detrimental to the public good if it stifles . . . research.

After one cuts through the smoke, this is the most important issue to be resolved about application of patent laws and doctrines to higher forms of life. These are new issues, not previously addressed, because application of patent laws to the seed/plant genetics supply industry is different than inanimate products. Ultimately, this question must be resolved equitably and fairly, taking into account the diverse interests, but in the final analysis decided on what is the best public policy. The ASTAs involvement has been for that purpose.

3 Bioethics—Impact of Proprietary Rights on Public Research Goals

A. David Kline
Iowa State University
Ames, Iowa

Universities have recently discovered the power, or at least the promised power, of proprietary rights. Intellectual property committees are busy formulating the administrative procedures for handling this special property. Patent officers are making the rounds to introduce themselves and their modern elixir that changes, not metal, but ideas into gold. This new found faith requires explanation in that the legal framework for exercising proprietary rights has been in place for some time. Why the current interest?

Once we have some insight into the explanatory issue, we need to ask whether the expected changes are good. More precisely, given that universities seriously pursue proprietary rights, how will this affect research goals and should we welcome these changes?

Though much of what follows will apply generally, my chief interest is in a narrowed domain. The universities in mind are the major public and private research universities, especially the land-grant universities. The central proprietary right of present interest is patenting, not copyrights, trademarks, or lesser forms of process or product protection. The intellectual domain to be considered is biotechnology, especially plant biotechnology.

PATENTS IN CONTEXT

To appreciate the role of patents in determining research goals one must locate patents as one of a cluster of mutually supporting factors driving research goals. There are serious structural problems in our national economy that the average person recognizes as inferior quality goods and services. We seem to have abandoned producing quality automobiles and consumer electronics, for example. That we produce more hamburgers and lawyers per capita than any other country is small reassurance. The problem can be put more analytically. During the 1970s the gross national product grew at rates significantly below that of other industrialized nations. Compared to any previous decade, our cost of living grew more rapidly. Growth of living standards became flat (Bok, 1982, p. 136–168).

Copyright © 1989 Crop Science Society of America, American Society of Agronomy, Soil Science Society of America, 677 S. Segoe Rd., Madison, WI 53711, USA. *Intellectual Property Rights Associated with Plants*, ASA Special Publication no. 52.

There are many causes of this malaise, but one is of particular relevance to the present discussion. Technological innovation has a significant effect on productivity and productivity has a significant effect on general economic health. During the 1970s and 1980s our rate of technological innovation has slipped. Industrial spending for research and development continues to drop.

Against this dark background, university research appears to be flourishing. American university scientists continue to produce a large volume of high quality research. Beginning in 1970, many in industry and government recognized that university research must be bridled to drive technological innovation more efficiently. Technology transfer has come to be seen as a top priority. Different institutional arrangements under the title of the university-industry complex have been and are being tried.

The university perspective on the complex is worth briefly noting. Immediately following World War II, interaction between universities and industries, which had been significant during the war, dropped sharply. But at present and at least for the last decade, universities have been under severe economic limitations. Real faculty salaries have been declining. Scientific research has become disturbingly expensive. The university science support infrastructure is cracking. Laboratories and instrumentation need dramatic improvements. It is not hard to see why the university sector is willing to listen to suggestions of an alliance with industry. In addition to their social responsibility to help the country, universities have hope for direct economic gain.

Derek Bok, president of Harvard University, in his book *Beyond the Ivory Tower*, endorses the new coziness between universities and industry (Bok, 1982, p. 142-2).

> In sum, the field of technology transfer offers all sorts of intriguing possibilities to stimulate technological innovation while reaping rewards with which to strengthen academic research. With hard work and a bit of luck, a university might conceivably contribute to the nation's prosperity in ways that could increase its own resources and thereby strengthen its research effort to make still further contributions to the economy—and so on in an endless synergy.

Patents, at least from the universities' perspective, fit nicely with the picture so far sketched. Patents will provide a formal mechanism for insuring that the supposed mutual benefit of the alliance is just that, a mutual benefit. Universities will not need to rely on the benevolence of industry to improve their laboratories and salaries of their faculty. Profitable patents rest on a licensing relationship, not benevolence.

We can now see in a rough way how university research will increasingly be shaped. Both the desires of industries and the self interest of universities will drive research to become more applied. Industry wants ideas that are at some level relevant to its markets. Universities do not want mere patents but rather patents that can garner handsome licensing fees. Again, if the plan is to succeed, research will have to be more market sensitive. The distinction between pure, fundamental, or basic research on the one hand and applied research on the other, despite its frequency of use, is far from clear. Nevertheless, it is fairly straightforward that much of what at present passes as basic research would not be the paradigm of the research appropriate for the university-industry complex. In the course of arguing for the new order, Derek Bok (1982, p. 152) chides faculty for looking down on applied work and failing to appreciate that there are problems of genuine intellectual merit in applied work. He clearly sees appropriate research moving in the applied direction.

In most cases research agendas will become more applied, but this will not always be the case. In some areas research will become less applied. The correct spirit of the university-industry compact is that university research should be industry relevant or market relevant (hereafter MR-research). It is important to realize that MR-research does not just involve a shift of emphasis on the basic to applied continuum. It also involves a shift in the goals of applied work. Not all applied work is acceptable. The research must be aimed at strengthening the economy through the traditional market mechanisms.

AN EXAMPLE—PLANT BREEDING

During the twentieth century the private plant breeding industry has applied political pressure to broaden its proprietary rights and influence the research agenda of public breeding programs. In the proprietary rights realm, the crucial legislation and court decisions include the Plant Patent Act of 1930, the Plant Variety Protection Act of 1970, the *Diamond v. Chakrabarty* Supreme Court ruling in 1980, and the 1985 *Ex Parte Hibberd* decision of the U.S. Board of Patent Appeals and Interferences. The legal history is one of continually broadening the domain of application for proprietary rights. At present, any plant or plant part is, in principle, eligible for patent protection.

Efforts to shape the research agenda are not so easily summarized. Most plants of major economic importance are not native to the USA. Beginning in the nineteenth century and continuing into the first quarter of the twentieth century the federal government, through the Department of Agriculture, the land-grant system of colleges, and state agricultural experiment stations, collected germplasm and distributed seeds of improved cultivars to farmers. The land-grant universities still provide finished cultivars.

The growing private seed industry not only faced intra-industry competition, but competition also came from farmers who saved their seed for planting and the public research system. Through hybridization techniques (provided by public research) and recent patent laws, farmers were virtually eliminated as serious competition. The real struggle has been to shape public research.

The federal government realized early on that it was desirable to have the agricultural sector populated with numerous, semiautonomous units. The economic infeasibility of these units carrying out their own research and development explains, in part, the origins of the public research system in support of agriculture. With the growth of the seed industry and the increasing potential for private profit, industry pressure has been for public research and development to be funneled through private industry rather than transferred more or less directly to farmers. The complaint from industry is interesting in that the charge is not for more applied work but for less applied work and more basic research. The land-grants are seen, at least in some of their activities, as unwelcome competitors. The Pound Report of the early 1970s and the more recent Winrock Report are noticeable signs of the pressure on land-grants to deemphasize extremely applied work and contribute to the national well being through what we have called MR-research. Frederick Buttel and Jill Belsky (1987, p. 37) nicely sum up the response of the public system.

> Lured by the "carrot" of increased federal and private funding of biotechnology and fearful of the "stick" of more scrutiny of the public agricultural research budget, . . . responses to the Winrock Report have been rapid by any standard. The resources of the LGU/SAES (Land-Grant Universities/State Agriculture Experiment Stations) system are being heavily diverted at the margin into biotechnology. LGU plant breeding programs, which represent the epitome of the highly applied research that is now out of favor, have been at the heart of recent shifts of LGU resources toward biotechnology. Plant breeding faculties are now hiring staff members who are molecular geneticists and tissue culture specialists rather than merely wheat, tomato, or sorghum breeders. The groundwork has thus been laid for further movement along the established trajectory of the division of labor between public and private agricultural research. The public research sector is increasingly devoting its resources to more fundamental research to be transferred to industry, and is deemphasizing highly applied research, especially that which leads to the development of products.

The shift in research agendas and the extension of proprietary rights are, of course, not independent of one another. They mutually support the dramatic changes that are taking place in plant breeding. The 1970 and 1980 extensions of proprietary rights helped make more areas of private plant breeding profitable, hence, the effort to squeeze out the overly applied product-oriented public research. Public research had to be redirected to fuel technological development rather than compete with private industry.[1]

The strengthening of proprietary rights has been good for universities as well. It makes the shift from applied to basic science very tempting. Product-oriented research was a service to the public, in particular, farmers and consumers. It was a use of resources that produced good will but little profit for universities. Pursuing MR-research has a chance of paying its own way and much more. Universities look enviously upon Stanford University and the University of California. Patents issued in 1980 based on the work of S. Cohen and H. Boyer for making "biologically functional molecular

[1] A commentator has made the interesting point that an alternative explanation to the Buttel and Belsky line is possible. ". . .biotech offers a new channel to new, more productive varieties and . . . more background work in understanding plant development etc. is required before that potential can be realized. That is, universities may have come to the restaffing choices entirely independently of industry pressure." This explanation does not seem to take the Pound and Winrock reports as important causal factors. The ultimate test will be whether the work of the biotechnologists is carried through to the applied level of traditional breeders or whether the private sector provides this finishing role.

chimaeras'' (a process for doing recombinant DNA work) have, to date, earned several million dollars under various licensing agreements. This is the model of how to help industry and to help oneself at the same time.

CRITICISM OF MARKET RELEVANT-RESEARCH

To ask about the effects of patenting on university research is an incomplete question. We need to be asking about the effects of an MR-research agenda, one formative element of which is the patent system. Obviously much university MR-research will be welcomed by industry and, more importantly from an ethical point of view, can be justified on the basis of its utility to society.

A broad critique of all MR-research does not look promising. Concerns must be focused more narrowly. Though it is not presently the case, it is not unreasonable to suspect that the MR-research agenda will grow to dominate the research agendas of universities. The proper interrogatory is: ''What reason is there to believe that knowledge produced from an MR-research agenda will always serve the public good *and* be sufficient for the public good?'' I shall mention two examples that are designed to show that knowledge based on MR-research is not inevitably in the public interest and develop two examples that show that MR-research is not sufficient for what we need to know, i.e., there is desirable knowledge that will be systematically excluded by the logic of a research agenda driven by MR-research. I first briefly consider several general critiques of university MR-research.

A number of issues have been raised in the literature, e.g., increasing secrecy within academic science, university science losing credibility, decreasing emphasis on basic science, and the unfair distribution of royalties. Each of these deserves lengthy treatment. Given that they take us from our main tack, I shall only offer a thought or two not found in the literature on the latter three issues.

In our times the assertions of members of few professional groups are regarded as credible. Academic scientists are a noticeable exception. An appreciation of the academic ethos and its emphasis on objectivity has produced considerable public respect for scientists.

American business does not receive such high esteem. The reasons for this are not hard to find. Ford Motor Company officials do not believe that the 1971 Pinto presents a greater fire hazard than was compatible with state-of-the-art automotive engineering. Also, RJR Nabisco does not believe that cigarette smoking is hazardous to one's health.

As university science makes deeper and deeper ties with corporate interests, the credibility of university science will likely suffer. Donald Kennedy, president of Stanford University, puts the point nicely (Kennedy, 1982; The Twentieth Century Fund, 1984).

> Fundamental research is still deeply respected in this country, much more so than most professions. Public confidence in its own objectivity is the best thing science has going for it; but without some care we may find one day that ''scientific objectivity'' has become an oxymoron, a self-contradiction like ''express mail'' or ''easy credit.''

I have mixed feelings on the credibility issue. On the one hand, I see science as a paradigm of rational inquiry and an activity that stands a chance of providing moorings in the pluralistic society. On the other hand, the loss of credibility may not be all bad. It is naive to forget that science takes place in a larger social and political context, and that that context affects the activity of science. Perhaps we will become less gullible and less likely to be satisfied by authorities. Perhaps we will become more scientifically literate and more able to evaluate scientific arguments on their own merits.

It is already too late to turn back. The politicalization of academic science is well underway. For example, the Council for Agricultural Science and Technology (CAST), a group that the Societies sponsoring our symposium officially support, recently issued a report, *Genetic Engineering in Food and Agriculture* (1986). The report aims to provide information to legislators, government officials, and the public on the nature and issues surrounding genetic engineering. Out of 47 pages, less than 2 pages are given to the risks of biotechnology. None of the 19 authors are political scientists, economists, or ethicists. Nowhere in the publication does CAST reveal its source of funding. The report argues for agricultural biotechnology on the grounds that the world needs more food in order to prevent starvation. After properly diagnosing the problem of hunger as political, the report reverts to calling for an increased food supply (CAST, 1986, p. 32). This is, of course, avoiding the real causes of hunger (Lappe' & Collins, 1979). Furthermore, it is transparently self-serving. Guatemala and the Philippines produce an abundance of food as hunger grows. (We will return to this issue later.) Nowhere does the report speak of the economic and political risks that biotechnology brings. Risks are understood totally in a narrow environmental sense. Even these are not taken up in a serious way. We are told that biotechnology is not importantly different from traditional breeding techniques. Nowhere is there a hint of the richness of the scientific debate and the fact that serious scientists have worries that cannot be dismissed with glib remarks. If the CAST report is an example of what scientific objectivity has come to, we are better off without it.[2]

A number of authors have registered their fear that increased MR-research will necessitate a decrease in basic research. I, of course, am sympathetic with this point, but it must be put carefully. As we have already seen, MR-research sometimes pulls in the more basic direction. Admittedly, this is not standardly the case. Nevertheless, the standard argument for basic research is far too weak. It gives the proponents of MR-research their key premise.

According to the standard argument, we need strong basic science programs because one never knows (one cannot know in advance) what basic

[2] In fairness to CAST, its recent leadership appears to have recognized some of the shortcomings of earlier studies. For example, *Long-Term Viability of U.S. Agriculture* (CAST, 1988), was composed by a heterogeneous task force that genuinely captures the diverse, and sometimes competing, interests in American agriculture. A study under preparation on the risks and benefits of antibiotic use in animals promises to be equally representative.

knowledge will be needed to solve some applied or technical problem. Therefore, to keep our applied work alive and vigorous, a strong and comprehensive program in the basic sciences is needed.

This argument from serendipity is less than compelling. It has at least two problems. Firstly, it supposes a simple linear influence of basic science on applied science. The actual situation is one of a complex feedback mechanism. On some occasions problems in applied science call for particular work in a basic area; on other occasions certain basic problems require prior solutions to applied issues. It is even the case that an area of applied science can be quite well developed prior to any basic or underlying science in the area. The theoretical or basic account may only come later. The relationship between the technology of steam engines and thermodynamics is a clear example here. Steam engines were invented and dramatically improved without the benefit of thermodynamics. My first point is simply that the relationship between applied and basic research is much more complex than the serendipity argument suggests.

Secondly, and more importantly, we should not grant that the end of science is solving applied problems. The end of science is scientific understanding. This occurs most profoundly on the basic level. This is the domain in which broad areas of science are united and our understanding is most complete.

Intellectual property committees debate the fair distribution of royalties from patents. Should the investigator receive 15 or 30%, or what? How much should the investigator's department and college receive? Should funds move to the wider university?

Let us take the question of fairness seriously. What does the investigator deserve, given his or her contribution? Remember that the investigator has benefited from 3000 yr of science, from a particular social structure that allows the flourishing of science, and from many dedicated teachers and fellow researchers. What part of the discovery is due to the merit of the investigator alone? Certain necessary components of most science, the calculus for example, took enormous genius to create. How much credit for today's work in physics is due to Newton and Liebniz? Recall that even Einstein by his own admission stood on the shoulders of a giant. How much does Fred Smith at State University deserve for the epicycle he adds to our knowledge?

My intuition is that in most cases he deserves very little. I do not want to quibble about percents, but 15 or 30% are wildly high. This shows something quite interesting. Our system of royalties is not a system of fairness or descent. It is really an incentive or motivational system. Two points follow from this: (i) We are attempting to induce scientists to behave in ways they are not naturally inclined to behave. We are introducing externalities into the problem set. Problems are not pursued on their scientific merits alone. (ii) Those that have warned that there will be pressure to do MR-research are correct. Derek Bok claims that those that do not want to pursue applied work will be left to follow their own scientific noses. This is not quite correct. Strong motivational influences are being introduced and perfected in order to shape the behavior of scientists. We should recognize it for what it is.

Let us return to the main task of specifying the detrimental changes in university research agendas that are likely to occur given the emphasis on patenting and the broader notion of MR-research. In the abstract two sorts of changes will occur: research that should not be done will be done and research that should be done will not be done.

I wish to concentrate on the latter category; but briefly, the first category is symbolized by herbicide resistant crops and bovine growth hormone. These cases have received considerable press. We could quibble with the details. I prefer to treat the cases as thought-experiments. They show that under our present economic structure there is nothing to prevent research that serves universities and agrichemical companies but does not serve the long-term interests of the public or farmers, i.e., it is research that does not aim at the common good. Similar cases will be played out again and again as the biotechnology business develops. Unfortunately in many cases, university MR-research will be fueling that development (Kline, 1989).

I wish to concentrate on the failure to do needed research because this point has not been widely recognized and because there is some chance of remedying the situation even without deep changes in our political economy. Two illustrations will be outlined.

Even from an egoistic moral philosophy we should be trying to strengthen the economy and social conditions in the Third World. Democracy will have a better chance and the people will be more vigorous trading partners under these circumstances. It is essential that peasant farmers not be driven from the land. More positively, it is essential that they be able to succeed as viable economic units. Agricultural colleges have a long history of trying to alleviate poverty in the Third World. These efforts should be intensified, not diminished.

We know that the Green Revolution has not been entirely a success. The benefits have been uneven, typically going to the larger landowners. In fact, in many cases the revolution has exacerbated inequality in the Third World. This disparity will likely be continued and accelerated with the biotechnology revolution (Kenney, 1983; Kenney & Buttel, 1985).

Corporations develop products for large markets. Agribusiness biotechnology is being turned to corn (*Zea mays* L.), wheat (*Triticum aestivum* L.), soybean [*Glycine max* (L.) Merr.], tomato (*Lycopersicon esculentum*)—the crops with large profit potentials. University research, if it is to be market relevant, will turn in the same direction. Peasant farmers have little need for machine processable tomatoes or high-yielding soybean. Since the peasant market is not an attractive one from a business point of view, the research they need will not get done. With the help from universities, more precisely with the lack of their help, the differences between rich and poor will grow.

The second illustration is admittedly somewhat anecdotal in that it is based on my experience of my institution. I have no data from which to generalize to other universities but, as we shall see, given the logic of the situation there is reason to think that the case can be extrapolated.

Who is responsible for carrying out the needed safety research? We cannot expect biotechnology firms to pursue the safety issue. There is no profit in such work. This is precisely the kind of cost businesses try to externalize, i.e., refer to the general society. I had an opportunity to put our question to the vice president for research of Calgene (Davis, CA). He informed me that his company was not pursuing safety research and that such research was appropriate for institutions such as Iowa State University.

This is not an implausible response given that state universities have obligations to the welfare of the broader society, and traditionally have been insulated from the vagaries and demands of the market. Well, what is Iowa State doing? It is important to realize that the institution has made an extraordinary commitment to biotechnology. A $35 million molecular biology building and $17 million plus in supplemental research funding are the most visible signs. There are internal shifts both in priorities and funding that will have dramatic effects on the future shape of the institution. The central question is, given these efforts, what sort of commitments have been made to a research agenda that addresses the safety issues? The answer is none. No faculty has research programs designed to address safety issues and there are no plans to hire faculty with the appropriate expertise.

My intention is not to "kiss and tell." Quite the contrary, Iowa State is the national leader in attempting to think through the broader issues surrounding biotechnology. The myopia with respect to safety is not a testimonial to the lack of thoughtful people but to the grip of the new paradigm.

We should have been able to predict this result. The emphasis on biotechnology at Iowa State has its roots in economic development and the university-industry bargain. Given the rationale of the bargain, safety research, as it does for Calgene, receives a low priority in the university. Safety research is not MR-research. It is unlikely to result in patents that are attractive to business. Put briefly, it is not the sort of research for which the bargain calls.

This result would be tolerable if there were other institutions to pursue safety questions. Agribusiness will not take on the task. It is universities that have the technical might, the credibility, and the commitment to serving the public appropriate for the task.

Matters are worse than the university simply not doing something that we would like it to do. The university, as a promoter of biotechnology, has an obligation to pursue safety research. The university is selling biotechnology to the public and must be held responsible for doing what is reasonable to insure the safety of biotechnology. A serious research agenda aimed at safety questions is not an unreasonable or extravagant demand.

SUMMARY

The consequences of patenting on university research goals in biotechnology must be understood in the broader context of the new emphasis on technology transfer. Based on the logic of the new emphasis, several detrimental consequences have been outlined including the failure of universities to pursue research agendas that they should be pursuing. The lack of research aimed at Third World agriculture and the lack of safety research are treated as paradigm cases. These anomalies are structural. They are not temporary aberrations that can be eliminated by good will and a few memos. They follow naturally from the new emphasis.

ACKNOWLEDGMENT

This work was supported by the Iowa State University Bioethics Committee, The Joyce Foundation and Northwest Area Foundation.

REFERENCES

Bok, D., 1982. Beyond the ivory tower. Harvard Univ. Press, Cambridge, MA.

Buttel, F.H., and J. Belsky. 1987. Biotechnology, plant breeding, and intellectual property: Social and ethical dimensions. Sci. Technol. Human Values 12:31–49.

Council for Agricultural Science and Technology. 1986. Genetic engineering in food and agriculture. Council for Agricultural Science and Technology. Ames, IA.

Council for Agricultural Science and Technology. 1988. Long-term viability of U.S. agriculture. Council for Agricultural Science and Technology. Ames, IA.

Kennedy, D. 1982. The social responsibility of innovation. Lecture at Yale Univ.

Kenney, M. 1983. Is biotechnology a blessing for the less developed nations? Mon. Rev. 34:10–19.

Kenney, M., and F. Buttel. 1985. Biotechnology: Prospects and dilemmas for third world developments. Dev. Change 16:61–91.

Kline, A.D. 1989. Biotechnology and the control of science, forthcoming. p. 1–17. *In* S. Gendel et al. (ed.) Agricultural bioethics. Iowa State Univ. Press, Ames, IA.

Lappe', F.M., and J. Collins. 1979. Food first: Beyond the myth of scarcity. Ballantine, New York.

The Twentieth Century Fund. 1984. The science business. Priority Press, New York.

U.S. Congress, Office of Technology Assessment. 1988. New developments in biotechnology—field-testing engineered organisms: Genetic and ecological issues. OTA-BA-350. U.S. Gov. Print. Office, Washington, DC.

4 A Congressional View on Proprietary Rights[1]

William A. Stiles, Jr.

House Agriculture Subcommittee on Department Operations, Research, and Foreign Agriculture Washington, DC

The 1985 Patent and Trademark Office (PTO) decision in *Ex Parte Hibberd* to allow the patenting of plants will affect a number of existing agricultural policies and statutes. It will also affect public institutions that are derived from or supportive of those policies and statutes. The patenting of plants will also cause changes in the food and agriculture sector. All of these areas need to be examined in order to maximize the utility of new inventions that may flow from this decision and thus maximize any subsequent public benefits. Such an examination is also needed in order to minimize the disruption caused by this decision and thus minimize any subsequent public cost. Areas where this examination reveals conflicts, or where a lack of data results in an inability to make a determination of benefits or costs, are areas that should be part of a public policy debate on the effects of the *Ex Parte Hibberd* decision.

Before examining the policy issues raised by the decision, some general observations and background need to be discussed. The agriculture production sector occupies a unique relationship with society. This relationship recognizes the sector's vital strategic contribution to society, and the biological and climatic uncertainties that affect agriculture. It also reflects a philosophy from our nation's past, one of strong public involvement in a more rural and agrarian society, that has been carried forward into our predominantly urban and industrialized society.

This relationship has resulted in a great range of public benefits and support for agriculture that are not found in other economic sectors: price and income supports; publicly funded research, extension, and other technical services; direct export subsidies; antitrust exemptions; and other benefits. In areas specific to this symposium, it has resulted in public programs of plant research and breeding as well as a germplasm exploration, collection, and maintenance program.

[1] Views expressed herein are the author's and do not represent the views of the committee or its members.

Copyright © 1989 Crop Science Society of America, American Society of Agronomy, Soil Science Society of America, 677 S. Segoe Rd., Madison, WI 53711, USA. *Intellectual Property Rights Associated with Plants*, ASA Special Publication no. 52.

This unique situation and the underlying reasons for it cause conflicts when elements of the broader society intersect with agriculture. Public policy decisions on these conflicts have frequently led to these elements being adapted when applied to agriculture in order to preserve this unique relationship.

An example of this adaptation has occurred in public policy decisions on proprietary rights as they apply to plants. The Plant Variety Protection Act (PVPA) (7 U.S.C. Sec. 2321, et. seq.), covering sexually reproduced plants, was an adaptation of patent rights enjoyed in other sectors on other inventions. This statute reflects the state of biological sciences at the time of enactment, combined with a continuation of the unique relationship between society and agriculture. This statute was a compromise between the demands of an increasingly sophisticated seed sector and the decision to preserve agriculture's public policy position.

The PVPA, compared to utility patent provisions, imposes less stringent conditions upon the award of breeder's rights, has a narrower scope of ownership, and less stringent requirements for conditions of uniqueness and for descriptions of the new cultivar (Lesser, 1986; OTA, 1984). In a concession to agriculture's unique policy position, the PVPA contains a crop exemption that allows a farmer to grow out protected seed for his/her use and to sell this seed to other farmers as long as it is done in compliance with state seed laws (7 U.S.C. Sec. 2543). It also contains a researcher exemption, explicitly allowing the research use of a protected cultivar (7 U.S.C. Sec. 2544). In addition, there is specific authority for the Secretary of Agriculture to force the open use of a protected cultivar "to insure an adequate supply of fiber, food, or feed" (7 U.S.C. Sec. 2404).

Now science and society, at least the broader economic interests of society, have again brought conflict into agriculture as a result of the *Ex Parte Hibberd* decision. This decision applies general utility patent law to plants without a recognition of the unique historical policy position that agriculture enjoys. One can argue that the policy change embodied in *Ex Parte Hibberd* more accurately reflects modern realities in the food and agriculture sector, and holds more promise for that sector. However, the existence of prior statutes, institutions, and public policy in this area presents a choice of either changing these prior actions to reflect this decision or adapting the patent decision to fit agriculture's unique relationship.

Some would argue that these broader public policy questions should not affect patent decisions; they are separate issues and should be addressed separately. Patent rights are part of a social contract in which a monopoly, anathema in our society, is granted on a temporary basis in order to achieve the social goal stated in the Constitution: "To promote the progress of science and the useful arts . . ." (Article 1, Sec. 8, Clause 8). The nature of this contract with society has been shown in policy decisions to preclude patents on certain inventions related to nuclear weapons (42 U.S.C. Sec. 2181). As proper acknowledgment of the costs of meeting the concerns of society, the government has extended the terms of drug patents to compensate inventory for marketing delays imposed by regulatory approval processes, fulfilling society's end of the bargain (35 U.S.C. Sec. 156).

The public policy issues raised by the *Ex Parte Hibberd* decision in this examination are grouped into four general categories. (i) The effect upon existing statutes, limited for this examination to the PVPA. (ii) The effect of the decision upon existing public activities and institutions. (iii) The effect upon the agriculture production sector and allied industries. (iv) The effect upon international activities, specifically the international debates on plant proprietary issues.

It should be noted that the second and third groups embody the same concerns raised about the effects of the PVPA during debate on the 1980 amendments to the PVPA (94 Stat. 3350). These constituted the major focus of a study of the PVPA that was requested by the Senate Agriculture Committee. In that 1985 study of the effects of the PVPA, the conclusions reached were limited by the lack of data and the inability to prove causal relationships between the changes in breeders' rights and changes in the agriculture sector (Butler & Marion, 1985). These limitations will undoubtedly apply to any efforts dealing with potential policy conflicts generated by the patenting of plants. Also, given the lack of experience with plant patents, this discussion must be viewed as an initial and speculative examination.

PLANT PATENTS AND THE PLANT VARIETY PROTECTION ACT

The first major area of policy examination is the effect of *Ex Parte Hibberd* upon the PVPA and its awarding of plant breeders' proprietary rights. These statutes run in parallel and, given the requirement of novelty contained in both PVPA and patent law, in theory duplicate claims under both statutes would not be allowed, since the granting of coverage under the first law would be disclosed and prevent the finding of novelty under the second law. However, in practice duplicate claims can be made, given differences in definitions and scope of proprietary coverage, and this issue will no doubt be the subject of court challenges.

Utility patents provide greater proprietary coverage and protection than the PVPA. The scope of these claims has not yet been refined, so it is possible that patent claims could reach back to parent lines, undisclosed traits in unimproved germplasm, and even identical traits in other plant species.

Patent protection also provides broader public disclosure of the plant cultivar than the PVPA and is far superior to some of the nonpatent options such as the use of trade secret provisions. This improved public disclosure, combined with the freedom to conduct noncommercial research on patented inventions, is the basis of the use of patent law to advance technology. However, PVPA, through its explicit research exemption and a narrower scope of coverage, has allowed greater researcher latitude in developing new cultivars from protected cultivars than will be allowed under patent law.

Utility patent law, unlike the PVPA, would require farmers to make royalty payments on second-generation seed produced by the farmer. The practicality of enforcing this provision, however, is difficult to envision.

In summary, the more comprehensive proprietary coverage under patent law leads one to conclude that utility patent use will predominate over PVPA use in coming years. This will require some adjustment, especially during the early period of plant patent coverage as PTO and court decisions adapt patent law to the unique conditions of plants and plant breeding.

However, these court decisions will be based on a narrow set of legal and scientific issues, will be decided on a case-by-case basis, and will not include the full range of policy issues involved nor consider the possible disruptions to the public agriculture programs and institutions. This situation adds to the need for a broader review of the policy issues associated with plant patenting.

PLANT PATENTS AND PUBLIC ACTIVITIES AND INSTITUTIONS

The second group of issues is potentially the most contentious since this includes activities and institutions developed in response to agriculture's unique relationship with society: public plant breeding and crop improvement efforts, the plant germplasm system, and the public agriculture research system.

One major issue in the area of public plant breeding is the potential restriction of the exchange of information and germplasm between researchers as patent law is applied to plants. An additional set of concerns centers on precluding the use of patented traits or processes in the development of new commercial cultivars during the life of a patent on a plant.

Some public breeders and researchers fear that the ability to secure patent rights to plant cultivars, and thus increase the commercial value of those cultivars, will inhibit the free exchange of information and germplasm. This issue was raised previously during the debates on the PVPA and was one of the subjects of the 1985 study of the seed industry (Butler & Marion, 1985). That study was unable to measure objectively the level of information and germplasm exchange but, based upon a survey, concluded that information flows from universities to the private sector probably increased while flows in the opposite direction probably decreased.

However, the scope of patent protection, much broader than under the PVPA, and the absence of an explicit research exemption has raised some new concerns. In a resolution approved by the National Wheat Improvement Committee (NWIC) in late 1986, a major research concern was clearly stated (I.B. Edwards, 1986, personal communication):

> This type of patent could also limit the free exchange of germplasm which is the driving force in plant improvement. Researchers would not want to receive or use patented germplasm fearing patent infringement even though their sources of genetic expression were completely different.

Added to the complexity of the current situation is the fact that many universities are now using patents to protect their inventions and generate needed research funds. This practice creates a situation where even exchanges between public researchers may be affected. In a 23 Dec. 1986 letter from the NWIC to the PTO, the committee expressed the opinion that, "attempts to protect and patent were by no means confined to the private sector and that university administrators were restricting their staff from the exchange of certain germplasm" (I.B. Edwards, 1986, personal communication).

However, the requirement that a patent be disclosed, and the provision that an invention can be the subject of a publication for up to 1 yr prior to patent application, is pointed to as an improvement to the sharing of the knowledge. Under the PVPA there are no disclosure requirements, and seeds deposited as a condition of PVPA protection are not publicly released. How plant patenting will affect information and germplasm exchange is unclear and deserves further study.

The critical question that needs to be asked in this discussion is, what is the purpose of the information and germplasm exchange? If it is to improve overall knowledge, then the patent system offers advantages. If it is to spur research toward the development of new cultivars, then the effect of plant patents may be mixed, due to a narrow definition of distance allowed between a patented cultivar and a subsequent cultivar that uses some of the traits of the original cultivar. Also, the broadened scope of the patent, now including seeds, plant parts, genes, and plant materials, is new territory for researchers who, according to the NWIC, may be reluctant to use patented germplasm. Also contributing to this concern is the absence of an explicit research exemption.

Much of this concern may stem from the untested nature of many of the patent definitions as they apply to plants. The scope and breadth of some of the more sweeping patent claims on plants have not yet been tested. With clearer definitions of some of the critical parameters of plant patent coverage, the public plant researcher's concerns may diminish.

One additional issue affecting researchers is the use of process patents to claim ownership of methods and techniques used in plant breeding. This become more important as biotechnology is used in this area. The Cohen-Boyer patent on an important gene-splicing method is an example of a process patent on a biotechnology technique. In plant breeding the Jones-Mangelsdorf patent for fertility restoration in male-sterile corn (*Zea mays* L.) breeding is another example. Whether these process patents, alone or in combination with patents on the resulting plant cultivar, will cause undue restrictions in scientific advances remains to be seen. While these patents are relatively old and advances in science have occurred, and even been accelerated by the discoveries, the widespread issuance of patents on new breeding methods needs to be evaluated.

The public plant germplasm collection and maintenance efforts, mainly within the operations of the National Plant Germplasm System (NPGS), are likely to be affected by the patent decision as well. At the start, however, a distinction must be made between a patented, finished cultivar and germplasm at a lesser state of improvement. Since a patent on an invention must show some practical or commercial utility, unimproved germplasm cannot be patented. Some of those seeing a threat to our public germplasm collection and maintenance efforts blur that distinction.

What will occur, however, is the patenting of improved traits that, depending upon the patent claims made and the eventual scope of the patent protection, may reach back to unimproved germplasm contained in relatives of a patented cultivar. This vertical reach across generations presents a difficult situation, since previously undescribed parts of the publicly held lines may unknowingly be swept into the category of private property. This concern is also shared by many of the public plant breeders and researchers who fear that utility patents on commercial cultivars will reach back to parent breeding lines and even unimproved germplasm. Sorting through this legal problem and keeping current with PTO decisions will require a major effort on the part of an already underfunded germplasm system.

The undefined coverage of plant patents raises a related issue. Without clearer delineation on the scope of plant patent claims, one can envision a patent claim approved for a genetic sequence in corn extended to that same sequence found in another feed grain. This horizontal reach across plant species could also sweep public germplasm into a protected status. Seeking to limit this confusing situation, the NWIC, in its 1986 resolution, limited its support for plant patents to "unique or novel genes not occurring naturally" (I.B. Edwards, 1986, personal communication). However, until the situation sorts itself out in court, confusion over coverage limits causes problems.

In addition, the germplasm system and the Department of Agriculture (USDA) are being swept into the middle of many of the difficult PTO decisions as they are asked to advise on the novelty and nonobviousness of patent applications. If a trait or cultivar claim is already known and described, it fails the test of novelty and nonobviousness, and cannot be granted a patent. This is a difficult task in any case, but with plants it gets more complicated since many of the accessions to the germplasm system were not described in the detailed way that patent applications are. Sorting these problems out is a major impediment to progress on plant patent issues and was one focus of a symposium held 4 May 1989 in Arlington, VA, titled "Plant Breeding and Biotechnology: Issues and Options for Intellectual Property Rights."

Finally, there are questions raised about the use of the germplasm system as a repository for patented cultivars. Under current patent interpretations, the deposition of the patented cultivar meets the requirement for disclosure of the invention. However, some of the current depositions are being made in repositories other than those in the NPGS, a practice that could frustrate attempts to centralize our information about plant germplasm and genetics. While the patented cultivar, upon certification and commercial sale, will find its way into the NPGS, the delay may cause problems. In addition, the day may soon come when disclosure requirements will be met by a description of the genetic sequence being patented, and not by actual deposition of plant material. Keeping pace with these developments and coordinating NPGS operations with other public repositories are major challenges.

For public research institutions, there is a special set of challenges that clearly exemplify the conflict generated by utility patents when they intersect agriculture's unique relationship with society. The public agriculture research and teaching institutions, the land-grant institutions, and the state agriculture experiment stations were established by laws passed in the late 1800s and early 1900s, which sought to support the agriculture sector by providing it with a solid technological base. This research and education subsidy to agriculture continues today with those institutions receiving about $1 billion annually from state governments. These institutions have viewed themselves as impartial researchers and educators, doing work in agriculture's interest as defined by the statutes establishing the land-grant system.

With changes in the patentability of the results of publicly funded research, universities have begun to use utility patents. In the land-grant university community, where much of the expertise in agriculture research resides and where most of the public plant breeders are located, plant patents will become another part of the university's proprietary income portfolio.

There is no question that the royalty income from patents to public universities is a benefit from patent law changes. For land-grant universities, the ability to patent plants and animals offers an avenue for removing the need for heavy taxpayer subsidization and offsets reductions in research funding experienced in recent years.

Another positive benefit from the use of plant patents by public institutions could be as a defensive strategy to keep plant lines and methods in the public domain. This strategy could become increasingly important during early periods of uncertainty over patent coverage as a means of protecting an institution or public breeder from seeing public, unprotected inventions from being swept into patent claims.

However, when we can no longer distinguish between a private univeristy, a private corporation, and a land-grant university by reviewing the patents that each holds, does the land-grant school still merit some special status in public policy and funding? What happens to the role of an impartial advisor to the farm community when a land-grant school is now engaging in the commercial development of farm inputs? Does this practice mean another decline in public plant breeding efforts as commercialization of research becomes more commonplace? All of these questions reveal policy issues needing attention.

PLANT PATENTS AND AGRICULTURE

The third major issue area is the effect of plant patenting upon the agriculture production sector, primarily the farmer. Also included are the industries that provide the inputs for farming, primarily the seed industry.

The agriculture production sector stands to benefit greatly from the accelerated improvements in plant cultivars that will flow from increased commercial activity in plant breeding. This increase will come from scientific advances, in areas such as biotechnology, coupled with the improved proprietary protection allowed by utility patents on plants.

These benefits, however, may entail some increased costs to farmers. And the use of plant patents may magnify changes occurring in the farm sector. In addition, the continuing concerns about the genetic diversity of our crop lines need to be reexamined.

One direct cost to the farmers is likely to be increased seed costs for patented cultivars. The cost increase due solely to the use of plant patents is difficult to examine, as pointed out in the 1985 study on the PVPA. This earlier study concluded that PVPA contributed to an increase in seed prices, and it can be reasonably expected that the *Ex Parte Hibberd* decision will have the same effect.

A paper by William Lesser on plant patents examined the possible effect of the patent law's effective elimination of the PVPA farmer's exemption on seed costs (Lesser, 1986). Since patent law would not allow a farmer to produce seeds from patented cultivars, increased seed costs could be expected for open-pollinated crops where farm production of seed is a common practice. The magnitude of the cost is very speculative and, without practical means of on-farm patent enforcement, it is doubtful that farmers can be prevented from continuing their current practices.

It may be, however, that a biological enforcement will occur as a result of commercial plant breeding that will end the practice of the farmer growing his/her own seed. Hybrid plant cultivars have biological protection against farmer seed production. As hybridization for improved performance is brought into other crops, this protection may expand. Biotechnology may accelerate hybrid development and may offer other means to control reproduction in subsequent generations. If widespread second-generation reproduction limitations are bred into crops, then some of the increased seed costs mentioned above may occur. In this case, however, the effect of patent law on any increase in cost would be mixed.

In the end, no farmer will pay a higher price for a cultivar if that outlay does not mean higher income. Assuming a continuation of a wide range of plant cultivar options to the farmer, increased seed costs from plant patents, or any source, will be a farmer's free choice based upon economic considerations.

However, if the range of options is narrowed, through changes in the seed industry or due to a decline in publicly released cultivars, the nature of the farmer's freedom to choose may change, altering the above neutral situation. In addition, if public plant breeding declines, the farmer will experience a decline in the public subsidy that public plant breeding represents, changing the indirect benefits that agriculture currently enjoys.

Whether plant patenting has any relationship to the issue of the genetic base of our crops remains to be seen. The 1985 PVPA study showed no relationship between passage of the PVPA and genetic diversity, mainly due to the lack of objective measures of genetic diversity (Butler & Marion, 1985). The NWIC 1986 resolution raised the possibility that a patent on a certain trait could limit research and development on the genetics of that trait, and thus limit the genetic diversity of the plant being bred to express the trait. This area shows the uncertainty spawned by the lack of specific delineations on the limits of plant proprietary rights under patent law.

The last major concern expressed in the farm production sector is the effect of plant patent law on vertical integration in agriculture. This concern, mostly expressed to date regarding animal patents and the animal agriculture sector, is that the distributors and processors of agricultural products may patent a cultivar, contract out the farming, and impose a crop purchase requirement upon the farmer. Thus, some fear vertical integration and concentration similar to that observed in the poultry industry could occur in agriculture crop operations.

It is unclear what effects plant patenting may have on this trend. There are certainly sound economic reasons for a commodity purchaser to vertically integrate, and patent protection on a plant cultivar may make that decision more attractive. The same integration, however, can be accomplished through currently available trade secret, forward contracting, and licensing arrangements. Whether plant patents have a role in this aspect of agriculture remains to be seen.

The effects of plant patents on the agriculture input industries, limited here to the seed industry, is the renewal of a concern raised during consideration of the 1980 PVPA amendments. At that time, some contended that the PVPA was leading to undue concentration in the seed industry and the takeover of smaller, independent seed companies. The 1985 study on the PVPA concluded that the passage of the PVPA in 1970 did contribute to the large number of mergers in the seed industry (Butler & Marion, 1985). The study, however, pointed out that the desire for diversification was a larger factor in the seed industry acquisitions.

Since that time, the increased interest in the commercialization of biotechnology has focused much attention on agricultural applications of modern biological technologies. Many companies already providing farm inputs, such as manufacturers of pesticides and animal drugs, are expanding their biotechnology research efforts, and some have purchased seed companies as a means of commercializing new product lines. Other companies in unrelated sectors have seen the opportunity as well and have followed suit. Whether the *Ex Parte Hibberd* decision will accelerate this, and lead to undue concentration in the seed industry, is unknown but bears close attention.

One additional structural issue that needs examination is found in the practical aspects of patenting and enforcement. Some have raised the point that patent litigation to defend and enforce patent claims is expensive and may be unaffordable to smaller firms. Given that we are in the early stages of the use of utility patents on plants, we can expect a great deal of litigation. Does the predominant use of utility patents for plants, and the expense of litigation, favor the larger companies over the smaller ones?

PLANT PATENTS AND INTERNATIONAL ACTIVITIES

The final category of policy consideration deals with the international component of plant patenting. The two major items involved here are: (i) how the *Ex Parte Hibberd* decision affects the position of the USA compared to other countries and (ii) how this decision affects the international debate over the ownership of plant genetic resources.

The USA's position to allow the patenting of plants generates some difficulties that need to be resolved with other countries who do not yet recognize or issue plant patents. The issue of allowing plant patents has generated a lively international debate, and it is expected that many countries will follow the lead of the USA in this area. The decision to allow plant patents will necessitate policy decisions, similar to those mentioned previously in examining differences between patent law and the PVPA. Specifically, this will require a reexamination of the breeder's rights agreements under the International Union for the Protection of New Varieties of Plants (UPOV) Convention of which the USA is a signatory country.

The final item is the effect the PTO decision has on the international germplasm debate. For some time, a debate on the ownership of and compensation for germplasm resources has been taking place at the Food and Agriculture Organization of the United Nations (FAO). In 1983, the FAO passed a resolution establishing an International Undertaking on Plant Genetic Resources, aimed at insuring the free exchange of plant genetic resources under the authority of the FAO.

The genesis of the debate that resulted in the passage of this resolution is the developing country contention that their plant genetic resources are collected by developed countries, who then claim ownership to the plant cultivars resulting from this germplasm. The developing countries feel that they are due some compensation for this exchange. The scope of the undertaking brought the objections of many developed countries, who argued that the germplasm resources in developing countries were undeveloped and only had value after the germplasm was improved, for the benefit of all people.

The developed countries argue that germplasm resources are the "common heritage of mankind" and should be freely shared. The obvious conflict between this position and the strengthened individual ownership of plant cultivars displayed in *Ex Parte Hibberd* has not been lost on participants in the international germplasm debate. Unless there is some resolution of this conflict, plant gene-poor developed countries risk losing access to plant gene-rich developing countries. This may be the policy issue needing the most immediate attention.

As can be seen by this cursory review of the changes that plant patenting may bring, conclusions are difficult to make. Even discussing the effects of plant patenting is a highly speculative undertaking, reflecting the uncertainties about how patent coverage will apply. Without a careful and thorough examination, this uncertainty will persist until PTO decisions, patent appeals, and court cases provide piecemeal clarification.

Continued uncertainly will inhibit progress in plant cultivar development. Corporate funds may be diverted from research to pay for patent challenges and defenses, an especially bleak prospect for smaller companies and newer companies seeking to enter the plant breeding area. Unresolved concerns about the scope of patent coverage may restrict research interest in patented cultivars and traits. Unresolved fears about the effects of expanded patent law application in agriculture will continue to fuel calls for patent moratoria.

In the agriculture sector, and among its supporting institutions, plant and animal patent availability has challenged past policy decisions. The production sector and the public research institutions need to redefine their respective goals and roles in modern society. The NPGS also needs to look to its future challenges, at a time when it is struggling to compensate for its past shortcomings.

The *Ex Parte Hibberd* decision provides both an opportunity and a need to review our plant breeding and germplasm activities, in both the public and the private sectors, both domestically and internationally. This review should be a comprehensive legal, scientific, and policy analysis involving as broad a range of viewpoints as possible. It should be initiated by USDA or Congress, with a mandate to conduct a broad review resulting in a very specific set of recommendations. It should have a clear goal of resolving uncertainties and conflicts that impede progress, and limit the benefits of plant patents to agriculture. It should also clearly state a set of national and international goals in the areas of plant breeding, germplasm, and genetic diversity.

As a means of guaranteeing that this examination occurs, continued availability of plant patents may have to be made conditional upon the completion of this review. This does not mean placing a moratorium on plant patents, but may involve an expiration of plant patent authority at the end of a set period of time unless the review has been completed.

Researchers use scientific principles while lawyers and judges use society's laws to state what can be done. It is the responsibility of selected governmental bodies, as custodians of the public interest, to decide what should be done. In the area of plant genetic resources and their commercial applications, this proper governmental involvement is long overdue.

SUMMARY

The extension of patent coverage to sexually reproduced plants in the *Ex Parte Hibberd* decision presents potential conflicts with existing plant proprietary rights. Plant patents may also generate conflict with existing public agricultural research, plant breeding, and germplasm programs. Because of uncertainties surrounding the scope of plant patent claims, the nature and extent of these conflicts is unclear, creating some confusion in the public and private sectors. This confused state begs for a comprehensive policy review.

REFERENCES

Butler, L.J., and B.W. Marion. 1985. The impacts of patent protection on the U.S. seed industry and public plant breeding. North Central Regional Res. Publ. 304. College of Agriculture and Life Sciences, Research Division, Univ. of Wisconsin, Madison, WI.

Lesser, W. 1986. Patenting seeds: What to expect. A.E. Res. 86-1. Cornell Univ. Agric. Exp. Stn., Ithaca, NY.

Office of Technology Assessment. 1984. Intellectual property law. p. 383–407. *In* Commercial biotechnology: An international analysis. OTA-BA-218. U.S. Gov. Print. Office, Washington, DC.

5 Specific Issues on Proprietary Rights

William H. Tallent

USDA-ARS
Washington, DC

I am pleased to join this group of speakers who represent diverse views on proprietary rights for plants. I will address some specific issues in this area that have caused us to reexamine our policy and procedures in the USDA. Although I am speaking as a representative of the department, I must caution that my views are strongly influenced by experience and responsibilities in the Agricultural Research Service (ARS), the primary research arm in the department. Other departmental agencies may have views that are quite different. In the Cooperative State Research Service, for example, their view is conditioned by the collective policies and practices of the 50 plus state agricultural experiment stations (SAES). Likewise, the attorneys in the Office of the General Counsel may have quite a different perspective on the use and value of proprietary rights for plants. Even in my own agency, there is some diversity of opinion about the wisdom and value of patenting plants. I shall try to represent all of these differing and perhaps conflicting views.

The ARS is in a pivotal position with regard to proprietary rights to plants. It has major responsibilities for preservation and dissemination of germplasm in the working and base collections in the National Plant Germplasm System (NPGS). These public collections are extremely important to U.S. agriculture and the research community, and contain unimproved germplasm of landraces and wild species as well as old and new cultivars. From the beginning, any bona fide user, breeder, or researchers has been able to obtain samples on request of any germplasm held in the collections.

The *Ex Parte Hibberd* ruling by the Patent and Trademark Office Board of Appeals in 1985 confirmed that plant inventions can be patented under the general patent law. Questions have been raised about the possible patenting of germplasm in the NPGS collections. The ARS has operated under the assumption that naturally occurring genes in public collections would not be patentable unless sufficiently altered to meet the inventiveness requirements under the law.

One of the issues to be addressed is the degree of change required to permit patent protection of germplasm. How much change is necessary to

Copyright © 1989 Crop Science Society of America, American Society of Agronomy, Soil Science Society of America, 677 S. Segoe Rd., Madison, WI 53711, USA. *Intellectual Property Rights Associated with Plants*, ASA Special Publication no. 52.

meet the requirements of novelty, unobviousness, and utility? What is the minimum distance between cultivars of similar genetic background that would justify protection? How do we measure this genetic difference?

As a part of the NPGS, the Germplasm Resources Information Network (GRIN) database of information has been developed, and provides an automated inventory of the many thousands of accessions along with passport and evaluation data about each accession. It is a current system that is updated by the curators of these collections and can be readily accessed by any user with a computer, modem, and telephone. There is a general perception that all plants or plant traits described in GRIN are not subject to patent. The presence of PVPA protected accessions and the possible presence of patented germplasm in the collections have prompted GRIN managers to flag accessions known to be protected. The ARS has developed a disclaimer statement for GRIN and curators to protect against inadvertent distribution of material not known to be protected.

Another proprietary rights issue of concern to ARS relates to the policies governing the release of new cultivars by federal agencies and the States. For many years the release of most field and horticultural crops was in accordance with the 1972 guidelines developed by the Experiment Stations Committee on Organization and Policy, and signed by each State Agricultural Experiment Station Director and the Administrator of ARS. The thrust of this policy was to develop and release improved cultivars to the public, to help assure free access to germplasm, and to provide a continuing source of seed of high quality and purity. Many of these cultivars were developed through cooperative breeding programs involving one or more SAES and ARS. In recent years, over 80% of the releases by ARS have been made jointly.

With the availability of protection under the Plant Variety Protection Act and the new interpretation of the patent laws, some SAES recognized an opportunity to earn income for support of their breeding programs through control over the release and distribution of new cultivars. The decision of some SAES to release cultivars under a licensing arrangement was a significant departure from traditional release practices. Both federal and state cooperators have been forced to reevaluate their policies on cultivar development and release, particularly when there are cooperative relationships with SAES that wish to patent or otherwise restrict distribution. The ARS has developed a policy that we believe is consistent with the intent of Congress, is in the best interest of U.S. agriculture, and is fair with our cooperators. Application of this policy, however, has been made difficult by the uneven policies on release and distribution of improved cultivars by state cooperators. Because of these policy differences, on several occasions ARS has had to negotiate complex ownership and release procedures for new cultivars developed cooperatively with individual SAES.

The ARS also cooperates with private industry and there has been a considerable number of joint releases of new cultivars. Over the last 8 yr, 15% of our cultivar releases have been jointly made with private industry. The mandate of the federal government to cooperate with industry has been recently strengthened by the Federal Technology Transfer Act of 1986 and by Executive Order 12591, which charges federal agencies to use the authorities provided in the act. The ARS has entered into several cooperative research and development agreements under the authority of that law; however, these agreements are deliberately vague concerning the issue of patents on plants. While it seems prudent to protect our options and maintain a degree of flexibility, we are fully committed to the policy of free exchange of germplasm worldwide.

Much of the interaction with industry relative to plant research will be in the general area of biotechnology rather than in traditional plant breeding. The biotech companies are generally pro-patent in their approach to protection of intellectual property rights. This is understandable because investment costs are high and competition is keen. Patents can provide protection from economic competition while markets are being established. The concern of ARS relates to the granting of patents that, because of lack of novelty or inventiveness, may actually serve as a disincentive to further research and development in the area protected by the patent. In some cases, a patent could seriously impair the further exchange of information or materials. We recognize that our conservative approach to plant patenting may be criticized at times by those who strongly support the use of patents to encourage technology development and transfer in all areas including plants and animals. However, as a major public agency supported entirely by public funds, ARS is obligated to act in the best interest of all the people rather than any particular segment of society. To the extent that patents may serve as a disincentive to the development and transfer of some particular technology, ARS is obligated to look at each request to patent on a case-by-case basis.

The ARS is not alone in this quest for the best solution to the issue of patents on plants. Individual commodity groups are also divided as to the most favorable position to take and how to protect new plants or plant products. The American Seed Trade Association represents several hundred seed companies, including some of the largest and some of the smallest. Even though these private companies all have a common objective to maximize their return on investment, they have not been able to agree on the use of utility patents on plant germplasm. The majority view seems to be that utility patents have value in some circumstances but that they should be used with restraint and only in carefully chosen situations. One minority view is opposed to the use of patents for almost all plant germplasm while an opposing minority group, which represents a number of seed companies claiming the major market share of the most extensively grown crops, is strongly supportive of the wide use of these patents. Who speaks for industry? With whom should ARS talk to ensure that its policies are supportive of the industries it serves? It sometimes is most difficult for a public agency to be fully responsive to such opposing viewpoints.

Finally, the department must also consider if and how the patenting of germplasm will impact international relations. Since the USA is heavily dependent on introduced germplasm for genetic diversity and crop productivity, the continued and unrestricted access to plant germplasm from other nations, particularly the developing nations, is of great importance to our long-term welfare. In recognition of this fact, the USA has maintained a policy of unrestricted free exchange of germplasm from its public collections. This policy does not extend to that germplasm held by public and private breeders or to the proprietary germplasm held by private companies. In a free enterprise system, the right to own and control property is protected by law. Therefore, improved germplasm held as private property by individuals, institutions, or companies cannot be made available by a public system without their express consent and approval. These policy differences are not well understood by those outside of the germplasm system in the USA or abroad. The department is sensitive to factors that may create a negative perception of U.S. policy and adversely impact free exchange of germplasm. The department will continue to have a special responsibility to advise the Department of State on germplasm and other international issues. If plant patenting seriously interferes with the flow of germplasm, the ultimate adverse impact may be much greater than incentive effects on research and revenues. The ARS wishes to assure that access to germplasm remains unrestricted between scientists, organizations, and nations.

Public agencies with broad responsibilities to collect, maintain, evaluate, enhance, and utilize germplasm for the public good have the responsibility to support the national policy of unrestricted exchange and to remove or minimize obstacles that interfere with the successful implementation of this policy. With the new interpretation of patent laws, ARS recently reviewed its cultivar release policy. While in the past ARS has had a policy not to patent plants, ARS now will consider patents on plants on a case-by-case basis when it is considered to be in the best interest of U.S. agriculture to do so. It is our intent to comply with the intent of Congress on this issue, and to maintain and strengthen our cooperative relations with SAES and private industry. In some cases, it may be desirable to get public use patents on new genetic material to assure that it will be freely available to the agricultural community. The ARS will continue to pursue policies and programs of research to develop and freely distribute useful information and technology to the benefit of all segments of agriculture and our society.

ACKNOWLEDGMENT

Dr. Paul Fitzgerald, Agricultural Science Advisor, Plant Germplasm, carefully reviewed and extensively revised this manuscript.

6 Intellectual Property Rights—Industry/University Agreements[1]

Victor L. Lechtenberg
Purdue University
West Lafayette, Indiana

The advent of utility patents on plants, animals, and other living materials is the most important change in the intellectual property landscape, as it pertains to plants, since the passage of the Plant Variety Protection Act (PVPA) of 1970. Many plant scientists are apprehensive about this change. This apprehension is based on uncertainty over the impact of patents on research and information exchange, and on technology development and transfer from universities and federal laboratories into products and services useful in maintaining a competitive agriculture of greatest benefit to society.

Such concern is not new. The passage of PVPA evoked similar discussion and apprehension among breeders and geneticists during the early 1970s. On balance, however, the impact of the act is perceived to have been positive. It may also be important to remember that the utility patent law has stood the test of time. There is little or no evidence that it has impeded research and technology development in those areas where it has been commonly used for many years. Most evidence is to the contrary. Technology transfer and adoption have been enhanced. The same may well turn out to be true for utility plant patents.

Regardless of the final verdict, concern over proprietary rights has already changed the nature of interactions between universities and private business. Unrestricted corporate support of university research has diminished. Increasingly, support has taken the form of contracts in which specific deliverables, including access to intellectual property, are expected by the company sponsor. The nature of these relationships is the subject of this paper.

JOINT UNIVERSITY-INDUSTRY RELATIONSHIPS

Several forces have been acting on both industry and the university to set in motion some of the changes reflected in emerging patent policy. Beginning in the late 1970s, U.S. industry, including agriculture, came under

[1]Contribution of the Purdue Univ. Agric. Exp. Stn., West Lafayette, IN 47907. Journal Paper no. 11 898.

Copyright © 1989 Crop Science Society of America, American Society of Agronomy, Soil Science Society of America, 677 S. Segoe Rd., Madison, WI 53711, USA. *Intellectual Property Rights Associated with Plants*, ASA Special Publication no. 52.

increasing pressure from foreign competitors. Our preeminent position in steel, automobile, and computer chip manufacturing eroded as we progressed into the 1980s. Agriculture has also faced unprecedented competition. Several studies—federal, state, and private—since 1980 have addressed competitiveness issues. One recent study (National Academy of Engineering, 1988) summarizes the technological aspects of international competitiveness. While the study deals primarily with nonagricultural arenas, the findings are applicable to agriculture, particularily as we attempt to improve existing agricultural products, develop new products, and apply high technology to crop improvement.

These studies have created awareness that technology plays an important role in competitiveness. There is also a sense that the system of technology development and transfer in the USA can be improved. Interactions and joint research efforts between federal laboratories, universities, and private industry have been encouraged by state and federal government. The Stevenson-Wydler Act[2] and the Federal Technology Transfer Act[3] encourage rapid transfer of basic research findings from federally funded projects into new industrial products and services. State governments have also become involved, using their universities as instruments for state economic and high-technology business development.

Many university-industry partnerships have been spawned by these state and federal efforts. To be successful these relationships must provide for appropriate protection of intellectual property, responsible licensing, and effective marketing by the industrial partner. Patenting and licensing is assumed to be the only effective method of technology transfer in high technology businesses, including agricultural biotechnology. In some cases, provision for such mechanisms of technology transfer has even been a prerequisite for acquiring state funds. As biotechnology tools are integrated into plant breeding and genetics, and to the extent that plant breeding and genetics become high tech, the same technology transfer instruments will be brought to bear.

BENEFITS OF UNIVERSITY-INDUSTRY PARTNERSHIPS

Several recent technological competitiveness reviews address the benefits from collaborations between industry and universities (National Research Council, 1987; National Academy of Engineering, 1988). These reviews conclude that joint research relationships between university and industrial partners benefit both parties.

Universities gain through:

Increase in Research Funding. Limited funds for agricultural research, especially in basic plant sciences, results in neglect of many important problems. In working with universities, companies are generally interested in very basic research—research that they cannot conduct as readily or that

---

[2] Stevenson-Wydler Technology Innovation Act. 1980. P.L. 96-480. Washington, DC.
[3] Federal Technology Transfer Act. 1986. P.L. 99-502. Washington, DC.

is only remotely related to a potential product line. In these efforts, university and industrial goals are similar—developing a better understanding of genes and gene products, plant biochemistry, gene expression, and response of plants to their environments.

Focus on Industrially Relevant Problems. Industrial involvement in university research helps assure that at least some funds are focused on those problems with the greatest potential for industrial payoff. Such focus is consistent with the goal of maintaining agricultural and industrial competitiveness.

Improved Technology Transfer. In some cases, results of university research fail to capture the attention of companies and entrepreneurs able to translate findings into products and services. Industrial involvement in the research program can help achieve effective translation of research results into products and services.

Employment Opportunities for Students. A large proportion of graduate students trained in university research programs find employment in companies. The opportunity to interact with company scientists during their research program inproves the students' understanding of the corporate research environment and establishes valuable contacts upon their entry into the job market.

Benefits that accrue to industry include:

More Ready Access to Research Findings. Universities and federal laboratories are recognized as the primary performers of basic research. When technology is evolving rapidly, access by industry to the findings of these basic research programs is valuable. Sponsorship of research provides an opportunity for early access, not only to protectable intellectual property but also, and perhaps more importantly, access to know-how (National Research Council, 1987). Know-how includes techniques, methods, developments, etc. that are usually associated with a state-of-the-art research program but that are not protectable except through secrecy and confidential agreements. Know-how is often difficult to acquire readily from published literature.

Contact with Leading Researchers. Through university partnerships companies can establish contacts with experts in the field that might be difficult to establish otherwise. In some situations these contacts can lead to important and productive consultancy relationships.

Student Contacts. Companies look to universities as the primary source for technically trained employees. Sponsored research relationships offer companies an opportunity to become acquainted with many students and postdoctorals. Students are usually not constrained by the experiences and prejudices of established investigators. Hence, students are a source of imaginative new approaches to solutions for problems.

Access to Special Research Services. Some companies, especially smaller and/or start-up companies, lack capacity for special research services such as amino acid or DNA sequencing. Collaborations with a university can make such services available, at least on a limited scale.

Access to Intellectual Property. Most industry-university collaborations provide companies some form of access to intellectual property. Some argue that such access is inappropriate in that licensees may acquire exclusive or limited exclusive access to technology developed at least partly with public funds. They may benefit disproportionately. This position must be weighed against the prospect that, without an opportunity for the licensee to recover development capital, the technology will go undeveloped. If this should happen, no one benefits from the public's investment.

INDUSTRY-UNIVERSITY AGREEMENTS

One of the more visible impacts of intellectual property rights on plant science research in universities has been the increasing private sector interest in licensing intellectual property and in supporting research under terms of a contractual agreement. Some of the terms and conditions in proposed agreements pose potential difficulties for universities, especially their agricultural experiment station components. The following comments address several of these potentially troublesome terms.

The Committee on Biotechnology of the Division of Agriculture, National Association of State Universities and Land-Grant Colleges, has published excellent guidelines for development of university-industry research contracts (Committee on Biotechnology, 1984). Other organizations (Government-University-Industry Roundtable, 1988) have also proposed model agreements. These model agreements and guidelines suggest language for terms in research agreements. Readers are referred to these documents for details.

In developing any research agreement, it is important for both parties to understand the goals each wishes to achieve. Universities usually wish to acquire funds to support research. They have the broad charge from society to engage in intellectual pursuits, advance the frontiers of knowledge, and teach students. To achieve these goals universities must be free to publish research findings in scientific literature. Companies, on the other hand, wish to acquire access to patentable intellectual property that can lead to saleable products. Their overall goal is to recover monies invested in developing these products and make a profit. These goals are not necessarily incompatible. Both parties, however, may need to compromise in order that both may reach their goals. It is also important to have one's institutional position clearly defined, as the terms of the agreement will simply be the vehicle by which this position is implemented.

Several common terms are occasionally difficult to negotiate in plant science and plant biotechnology research agreements. These are not presented in any priority order. Comments regarding these terms are strictly the opinions of the author.

Research Exemptions. Universities should always retain ownership. They should also retain the right to use all intellectual property developed during a project, whether protectable or not, in their research and education programs. If possible, they should also retain the right to grant other universities and research institutions similar use.

Publication Rights. Universities must retain full control over publication decisions. Opportunities may be provided for sponsors to review manuscripts in draft form, offer comments, and assure that the manuscript does not contain confidential information. Sponsors may also be granted the right to request a delay in publication. In the authors opinion, such delay should not exceed 6 months. Sponsors should not be granted approval authority over publications. In view of the university's need to publish, trade secrets are not a viable option for protecting intellectual property.

Project Control. Sponsors may want the opportunity to suggest modifications and changes during a project. Such input can usually be accommodated. However, the sponsor should not be allowed control over the project, or over the scientists and technicians engaged in the research. The university must, however, agree to use its best efforts to carry out the project as proposed.

Exclusivity of Funding. Sponsors may want assurance that they are the only source of funds for an individual scientist, or that they will have access to all intellectual property developed in a particular research area. These terms are generally unacceptable to universities. They inevitably lead to constraints on exchange of research information, constrain cooperative efforts, and restrict the principal investigator's ability to seek funds from other sources, including federal grants. An acceptable alternative is to grant the sponsor an option to fund additional projects proposed by the principal investigator in the same subject area. If the sponsor declines, the investigator is free to seek funds from other sources. It is especially critical that the research project be specifically and narrowly defined. Proposals to cover broad, general areas of investigation should be avoided.

Intellectual Property Licenses. Intellectual property owned by the university can be licensed to the research sponsor. License provisions can grant exclusive, limited exclusive, or nonexclusive use for commercial purposes. Plant biotechnology, and plant genetics and breeding projects may produce intellectual property that is patentable under the utility patent law, or protectable under the PVPA. In either case, license provisions should contain appropriate research exemptions. Nonexclusive licenses or mandatory sublicense clauses can assure that products are accessible by large numbers of users.

When the intellectual property consists of a saleable product, such as a cultivar or line, and the sponsor's funding represented only a portion of the total development effort, the university may insist on a time limit for any period of exclusivity. Three to 5 yr may be appropriate. When patentable intellectual property, such as a gene, is licensed, the university may wish to retain the right to use such property in its own plant breeding and genetics program, including the right to release cultivars and lines in its traditional manner.

March-In Rights. Sponsors must agree to use best efforts to bring licensed technology to commercial fruition as rapidly as possible. If they fail the university should have the right to withdraw the license and issue to third parties. March-in provisions also prevent a sponsor from withholding technology that might compromise the market position of a current but perhaps inferior product.

Confidential Information. It should be understood that no information developed or generated as part of the research project is to be confidential. Any confidential information should consist solely of information belonging to the sponsor at the time the project started, or developed independently by the sponsor. Such information should be provided to the university researchers only if necessary to carry out the project. Such information should be disclosed only to the principal investigator. Confidential information may be critical to the proprietary position of the sponsor, and must be protected and kept confidential by the university (Zschoche, 1986).

Project Scope. Precisely defining the scope of the project is essential to avoid future disagreements over funding and property rights positions. As stated earlier, broad, generic descriptions of a general scientific area should be avoided. Project descriptions should be precise and constrained to specific species, systems, pathways, etc.

Funding Schedule. University business offices will usually insist that funding occur in advance of the expenditures. Suitable payment schedules can usually be negotiated. Provisions should commit the sponsor to all obligations that cannot be cancelled should the project be terminated for any reason.

Consulting Arrangements. One of the most effective forms of technology transfer is through consulting arrangements. Consulting agreements usually contain property rights clauses in which the company claims ownership to all intellectual property developed under the consultancy. Conflicts of interest can arise when a scientist enters into a consulting relationship with a company sponsoring his university research. When companies will both sponsor research and engage the principal investigators as consultants, the agreements should clearly address the ownership of intellectual property. A workable option is for the university to claim all property developed within the scope of the sponsored research project, regardless of whether developed as part of a university research program or under the consultancy.

SUMMARY

Concern over international competitiveness has caused federal and state governments to encourage joint research relationships between universities and industry. As these joint efforts have evolved, experience has been gained in developing contracts that protect the goals and interests of both parties. Since these goals differ significantly in many respects, compromise is essential. Joint efforts can work successfully if both parties agree to carefully worded terms and conditions governing property rights, license arrangements, and a number of other issues.

REFERENCES

Committee on Biotechnology, Division of Agriculture. 1984. Guidelines for development of university/industry research contract. p. 50–65. *In* Emerging biotechnologies in agriculture: Issues and policies. Progress Rep. III. National Assoc. of State Universities and Land-Grant Colleges, Washington, DC.

Government-Industry-University Roundtable. 1988. Simplified and standardized model agreements for university-industry cooperative research. Industrial Res. Inst. Natl. Academy Press, Washington, DC.

National Academy of Engineering. 1988. The technological dimensions of international competitiveness. Natl. Academy Press, Washington,DC.

National Research Council, Board of Agricultural. 1987. Agricultural biotechnology—Strategies for national competitiveness. National Academy Press, Washington,DC.

Zschoche, W.C. 1986. Technology transfer issues: An industry perspective. p. 157–166. *In* J. Rhodes (ed.) Agricultural science policy in transition. Agric. Res. Inst., Bethesda, MD.

7 Proprietary Rights—Private Industry

Roger A. Kleese
Molecular Genetics, Inc.
Minnetonka, Minnesota

My task is to address specific issues on proprietary rights as a representative of private industry. However, there is not a consensus private industry position on proprietary rights; rather, the American Seed Trade Association (ASTA) paper on "Patenting and Plant Variety Protection for the Seed Industry" (ASTA, 1988a) presents sharply contrasting viewpoints. The diversity is not surprising. No two companies in the seed industry come with a common set of experiences or business situations. Some are independent, others are owned by another firm; some are large, others small; some have research and others do not; some are profitable, others are not. They vary in the species marketed and strategies for marketing those species. It is not possible to give a balanced industry perspective.

I will attempt to speak on a limited number of issues, but because of my own bias, I will not attempt to argue objectively for and against both sides of an issue. I speak from my own experience and current business situation. Therefore, it is important that I describe briefly for you something about our firm.

Molecular Genetics, Inc. (MGI) was put together in 1981 with the belief that some of the new technologies layered on a plant breeding base could be used to create unique and useful seed products. We attempted then, as we do today, to make decisions about the crops we work on and the products we develop based upon the resources available (finances, germplasm, people, technology), the feasibility of making the product, and the market potential. Protecting intellectual property was not an important issue in 1981. We did see health care–oriented biotechnical firms active in filing patents, and, of course, certain of those had already issued, but none had issued for genetically modified plants.

The only species we chose to work with was corn (*Zea mays* L.), and that decision was driven primarily by market opportunity. That market and its potential profitability were attractive because hybrids were the product. In contrast, our reason up to this point for not working with other species,

---

Copyright © 1989 Crop Science Society of America, American Society of Agronomy, Soil Science Society of America, 677 S. Segoe Rd., Madison, WI 53711, USA. *Intellectual Property Rights Associated with Plants*, ASA Special Publication no. 52.

particularly soybeans [*Glycine max* (L.) Merr.] (even though the market is large), is that cultivars, not hybrids, are the product. Gross margins, repeat sales, and protection of germplasm are all issues to us.

Molecular Genetics, Inc. has adopted the policy of protecting our inventions with U.S. patents and trade secrets. It is not always easy to sort out which inventions are worthy of patent protection, and I believe the industry will find a number of patents that have no value. Since the technology is new, the tendency is to imagine the current technical hurdle as the most difficult in an entire product development plan. Thus, researchers may file for a patent before finding there is yet another major hurdle more difficult than the first. We have faced this problem but have tried to refrain from spending time writing unnecessary patents.

While part of our incentive to patent has been to publicly lay claim to what we have done, recognition has not been a driving force in our filing patents. Instead, we have truly believed (perhaps with some naivety) we had something of value and wanted to be sure we held onto it. We have sought patents for those inventions where the technology of the invention would eventually be commercially utilized. This practice has been influenced by a partner's concern for protection also. It has been our assumption that a U.S. patent provided maximum protection.

Inventions that enable us to do our research more effectively generally will not be patented but will be kept in-house as trade secrets. Our reason for this is the continual evolution of the technology and our own confidence in our ability to improve the systems with which we work.

To date we hold two U.S. patents, one for tryptophan overproduction (Hibberd et al., 1986) and one for herbicide resistance (Anderson & Hibberd, 1988). These patents cover unique and useful traits, and we intend to continue filing similar utility patents.

When both patents were filed in 1984, there were no U.S. patents on genetically modified plants. We were advised that both inventions should fall within the scope of U.S. patent law and that we should proceed with the filings. Action on the tryptophan patent moved more quickly than that on the herbicide resistance, and it eventually became the test case. The examiner in the Patent and Trademark Office (PTO) rejected the application on the basis that it did not fall within the subject matter as defined by the U.S. patent act. He argued that the Plant Patent Act of 1930 and the Plant Variety Protection Act (PVPA) in 1970 covered the exclusive forms of patentable protection for plant life. We disagreed.

We appealed to the PTO's Board of Patent Appeals and Interferences. The board concluded there was no intent that the plant specific acts be exclusive, and they unanimously reversed all of the examiner's rejections.

This landmark case established the legal precedent for the patentability of genetically modified plant material, a provision that had always existed in U.S. patent law. On 15 Apr. 1986, a U.S. patent issued for Tryptophan Overproducer Mutants of Cereal Crops (Hibberd et al., 1986). I want to describe the breadth of the patent, but first it is necessary to explain how these mutants were obtained. Tryptophan analogs were used to select for cultures deregulated for the limiting enzyme, anthranilate synthase. The invention covers "novel plant genotypes or lines which produce increased free tryptophan levels and are resistant to inhibition by analogs which normally inhibit the activity of the enzyme anthranilate synthase. . .to genes encoding such enzymes, and to processes for utilizing these novel genes and enzymes." The "products of the invention include plants, plant tissues, and seeds which exhibit resistance to such analogs and overproduce tryptophan resulting from expression of genes encoding analog resistant anthranilate synthase enzyme" (Hibberd et al., 1986). Processes and products are both covered.

On 2 Aug. 1988, a U.S. patent issued for Herbicide Resistance in Plants (Anderson & Hibberd, 1988). This, too, is a utility patent covering products and processes relevant to herbicide resistance. Today, a growing number of U.S. patents have issued, and the expectation is that this trend will continue. The U.S. patents are to be a part of doing business in the seed industry.

SOCIAL ISSUES AND U.S. PATENTS

Since obtaining U.S. patents on genetically modified plant materials represents a change for the industry, it is reasonable to ask whether such a change is in the public's best interest. Are there social issues relevant to seeking U.S. patents on genetically modified plant material? The extreme position is that all germplasm should be in the public domain. However, common practice in the seed corn industry does not support this position. Proprietary inbred lines are not released to the public. They represent unique genetic material that has been altered by the seed company whose right it is to keep such lines within the firm. Because corn is sold as hybrids, the improvements can be commercialized without releasing the lines to the public. Thus, it is common to claim particular improved genetic materials as private property, and it is generally understood that doing so is the right of the inventor.

Do U.S. patents alter this assumption in some way? By definition of a U.S. patent, "A patent is a legal monopoly granted by the U.S. government to an inventor which permits the inventor to prevent anyone else from making, using, or selling his or her invention for 17 years. The inventor can practice the patented invention exclusively, can license the right to practice the invention for royalties or can sell the patent outright" (Woessner, 1986). The words *legal monopoly* are a demanding part of this definition.

How is it that patents are intended to be in the public interest? "The patent system was created to help achieve various socially desirable goals. By providing an inventor with an exclusive right to make, use, or sell an invention for a limited period of time, it serves to reward an inventor for the time and effort expended in developing the invention, and thereby encourage further creative efforts. Since most new inventions have uncertain commercial value, the patent system provides a degree of protection from competition for a limited period of time, thus encouraging investment in new technology. The patent system encourages inventors to make their inventions known rather than to maintain them in a state of secrecy. This increases the amount of technological knowledge available to the public'' (Minnesota Small Business Assistance Office and Merchant, Gould, Edell, Welter and Schmidt, P.A., 1986).

In light of the above it is clear we have the legal right to patent genetically modified plant materials. Furthermore, we believe the patent system is intended to achieve socially desirable goals, and that obtaining U.S. patents is at the same time consistent with working toward those goals and our business goals.

Patents bring to the seed industry a legal definition of the claims of an invention. Without this definition it would be extremely difficult to define the limits of many inventions, particularly where brand new product ideas are being patented. In addition, we are convinced that patents will stimulate creative research.

CONCERN ABOUT WHAT WILL BE PATENTED

Molecular Genetics, Inc. has no formal policy for deciding which inventions to patent. Decisions are made on a case-by-case basis and always involve such criteria as market value of the product, ability of others to duplicate the invention, and our ability to protect it. Generally, we will not limit patents to particular kinds of technology.

I recognize that our criteria for deciding what to patent are vague. Since patenting in the seed industry is new, it is not clear what will be patented either as improved versions of existing products or as totally unique products. There was early concern about attempts to patent the obvious, for example, common traits such as standability. Those visions have been dispelled as people learn more about what patents are and the criteria for obtaining a patent. The criteria used by the patent office are nonobviousness, uniqueness, and utility. While examiners attempt to use these criteria rigorously and consistently, what is obvious to one of us is not to another, and what is unique to one may not be to another. It will be helpful to learn more about U.S. patents and the procedures used by the patent office in examining an application. As new patents issue, we are beginning to see how the U.S. patent office is using these criteria in the seed industry.

ACCESS TO GERMPLASM

The ASTA ''Statement of Basic Concepts and Principles for Patent and PVP Rights for the Seed Industry'' holds to the position that a compulsory licensing system should be established when the ''scope of a product or process patent is so broad as to encompass within its scope varieties or parts of plants not yet developed'' (ASTA, 1988b). We are opposed to this position because it significantly reduces the inventor's competitive advantage and it does not stimulate research in the industry. A patent is a legal monopoly. That is how patents were intended and how they have been used in other industries. I am not aware of this generally working to the detriment of those industries. It is interesting to think about the breadth and value of U.S. corn germplasm had we operated with patents over the past 50 yr. Perhaps a handful of inbreds would not have dominated the industry. Maybe there would have been a broader base with more heterotic groups and more options for good hybrids.

Access to germplasm has always been an issue in the seed industry. Germplasm, the necessary raw material to build a sales and marketing operation, has been accessed and used directly by the seed industry, many seed companies having become profitable businesses in the process.

Since all germplasm can be improved, some companies have chosen to invest in research with the objective of further improving their lines and populations. The business incentive is the development of unique proprietary lines and the security of not being totally dependent upon lines developed outside the company.

A major source of germplasm for these companies has been the breeding programs at experiment stations of land-grant universities and the USDA. Resources to support these programs have been reduced significantly in the past 15 to 20 yr; the results are fewer programs and less elite germplasm being released. The need to obtain research funds for these programs has led to tighter restrictions on the release of new sources of germplasm. Interestingly, the PVPA of 1970 has provided a useful mechanism for realizing value from these public sources of germplasm. In some instances release, licensing, and royalties are similar to those used by private industry.

Increasing competition for elite material is causing the seed industry to change. I do not believe all seed companies of the next 10 to 15 yr will have research organizations, but they will need to work harder at accessing material from productive research programs. Whereas in the past most seed companies focused on germplasm, various forms of intellectual property will become increasingly important. Even large, fully integrated, research-based companies will become more active in accessing intellectual property. Technical know-how necessary for product development will be licensed with or without germplasm. Germplasm may in some cases simply be a gene or a plasmid.

The smaller the research organization, the more they may be faced with acquiring rights to intellectual property. A firm without a research group must acquire from others all of its intellectual property. Many businesses routinely do this even in industries where patenting is used extensively.

SUMMARY

At issue is whether everyone will have access to any piece of intellectual property. I believe that will not be the case and businesses will have to be flexible enough to alter their business plans. For example, we hold a patent for herbicide resistance (Anderson & Hibberd, 1988). The work leading up to this patent was done under contract with American Cyanamid (Wayne,

NJ). The terms of the contract granted them the right to market the products of the technology with only a royalty flowing back to us. We agreed to these terms because we needed to finance our research and to develop technology. It was a business decision that made sense at that time. Today, we would prefer to be taking the product to market ourselves but we are not. We have considered the impact of this decision and adjusted our business plan accordingly.

The seed industry is always changing. It is important to understand the forces involved, and today, proprietary rights are one of them. The demand is not simply to respond to, but also to anticipate and responsibly direct those changes. Communication within the industry is a necessary step for this to take place.

REFERENCES

American Seed Trade Association. 1988a. Patenting and plant variety protection for the seed industry. ASTA, Washington, DC.

American Seed Trade Association. 1988b. Statement of basic concepts and principles for patent and PVP rights for the seed industry. ASTA, Washington, DC.

Anderson, P.C., and K.A. Hibberd. 1988. Herbicide resistance in plants. U.S. Patent no. 4,761,373. Patent and Trademark Office, Washington, DC.

Hibberd, K.A., P.C. Anderson, and M. Barker. 1986. Tryptophan overproducer mutants of cereal crops. U.S. Patent no. 4,518,847. Patent and Trademark Office, Washington, DC.

Minnesota Small Business Assistance Office and Merchant, Gould, Edell, Welter and Schmidt, P.A. 1986. A guide to intellectual property. Minnesota Dep. of Energy and Economic Development, St. Paul, MN.

Woessner, W.D. 1986. Where do patents come from? The Minn. Chemist 38:3–6.

SECTION II

An Evaluation of the Issues, Challenges, and Opportunities Related to Plant Patenting

Part A. Background on Biological and Statutory Bases of Intellectual Property Rights Related to Plants

PREFACE

The first section of this special publication presented a wide array of experiences and opinions about policies and practices related to proprietary aspects of plant germplasm. In Part A of the second section of the special publication, papers are presented that give an authoritative and comprehensive review of the main biological and legal issues of ownership and proprietary protection of plants. These papers provided background information for the discussants at a workshop held at Anaheim, CA, 31 Jan. to 3 Feb. 1989. The workshop was attended by about 80 invited specialists who represented the various disciplines of agriculture, biology, economics, social science, ecology, and law. Part B includes the summary report developed by the combined deliberations of the workshop membership. It also includes a presentation of the methods used at the workshop. The workshop included five discussion topics, each addressed by multidisciplinary groups of specialists. The Workshop Report is a summary developed from the reports of the five discussion work groups and from general discussion by the whole group of specialists. The Work Group Reports of the Workshop are available as a separate publication from the Crop Science Society of America (CSSA) Headquarters Office.

This evaluation was organized by CSSA, along with its sister societies, the American Society of Agronomy (ASA) and the Soil Science Society of America (SSSA), and cosponsored by the American Agricultural Economics Association (AAES) and the American Society for Horticultural Science (ASHS).

Financial support was provided by the USDA, ASA, CSSA, SSSA, and the organizations sponsoring participants at the workshop.

Calvin O. Qualset and Donald N. Duvick cochaired a steering committee that developed the concept and detailed format for the evaluation. The steering committee, included Robert F Barnes (ASA, CSSA, SSSA), John H. Barton (Stanford University Law School), Richard L. Lower (ASHS and University of Wisconsin), Richard K. Perrin (AAEA and North Carolina State University), Lisa J. Raines (Industrial Biotechnology Association), Henry L. Shands (USDA-ARS), Howard Silverstein (USDA, Office of General Counsel), Sidney B. Williams, Jr. (UpJohn Company), and Paul J. Fitzgerald, ex officio (USDA-ARS).

The editors compliment the authors of chapters who completed their assignments with short lead time.

Laura Meagher (Rutgers University) and John Barton were extremely generous of their time in developing the summary document, Part B.

We especially want to thank Orville G. Bentley, former Assistant Secretary for Science and Education, USDA, and Paul Fitzgerald and William Tallent (USDA-ARS) for their great interest and administrative support to this project.

Editors

John H. Barton, *Stanford Law School, Stanford, California*

Calvin O. Qualset, *University of California, Davis, California*

Donald N. Duvick, *Pioneer Hi-Bred International, Inc., Des Moines, Iowa*

Robert F Barnes, *American Society of Agronomy, Madison, Wisconsin*

1 Plant Breeding, Crop Cultivars, and the Nature of Genetic Variability

Fredrick A. Bliss
University of California
Davis, California

PLANT BREEDING

Qualifications of the Plant Breeder

Genetics is the foundation of plant breeding, but plant breeding is more than applied genetics. The plant breeder should possess in-depth knowledge of plant biology and have formal training in classical genetics, population genetics, biometrical and quantitative genetics, cytogenetics, and molecular and cellular genetics, as well as crop evolutionary processes. An appreciation of production agriculture will aid the breeder in developing and executing a relevant research and development program. The experimental nature of plant breeding requires knowledge of statistical methods. Since plant pests and diseases are often limiting factors, resistance breeding is a common objective that requires an understanding of plant pathology and entomology. The complex nature of plant traits requires a working knowledge of biochemistry and plant physiology to effectively improve plant productivity through the use of physiological genetics. Because the approach to modern plant breeding is multidisciplinary, often the plant breeder is the integrator for a team effort.

Steps in the Breeding Process

Plant breeding begins with analysis and definition of the problem, establishment of program goals, and definition of specific objectives. The next step is selection of germplasm that possesses the traits needed to meet the program goals. This germplasm may be used either as potential new cultivars or a parents for crosses to produce genetically variable populations. The working populations provide important raw materials for selection. Good materials are added to and poor ones discarded from this gene pool.

Choice of breeding or selection methods depends on the mode of plant reproduction, the heritability of the trait(s) being improved, and the type

Copyright © 1989 Crop Science Society of America, American Society of Agronomy, Soil Science Society of America, 677 S. Segoe Rd., Madison, WI 53711, USA. *Intellectual Property Rights Associated with Plants*, ASA Special Publication no. 52.

of cultivar used commercially (e.g., F_1 hybrid cultivar, pureline cultivar, etc.). For highly heritable traits, choice of superior individual plants evaluated at a single location will be effective, whereas for traits with low heritability, selection should be based on mean values obtained from replicated evaluations of families of related plants. Popular selection methods commonly include pedigree selection, modified pedigree selection, mass selection, and recurrent selection.

The complexity of inheritance influences choice of the breeding method. Backcross breeding is used to transfer one or a few favorable genes for a highly heritable trait into a desirable cultivar. This approach has been used extensively for breeding disease-resistant cultivars. Recurrent selection is used to improve quantitatively inherited traits controlled by numerous genes. The use of recurrent selection in self-pollinating crops depends on the ease of pollination, the frequency of successful hybrids from each pollination, and the number of hybrid offspring from each successful cross.

Each breeding program should include a periodic, objective evaluation of the efficiency of the breeding procedure. Evaluation criteria vary depending on the goal and objectives, but should include gain from selection per year based on comparisons to an appropriate standard, overall value of the advanced breeding lines, and number of successful cultivars produced per unit of input (e.g., per year, per dollar expended, etc.)

Promising advanced breeding lines are thoroughly tested and compared to appropriate standards in environments representative of the commercial target area(s) for at least 3 yr. The best lines are candidates for new commercial cultivars; those still deficient in a few traits are used as parents to produce new populations for further selection.

These processes which lead to the final step of marketing and distribution, usually take from 7 to 10 yr from the time the first cross was made. Therefore, development of new cultivars is a time-consuming process that requires precise forward planning, efficient use of resources, and a minimum of changes in direction.

An attractive feature of new biotechnology is the potential for reducing the time required to produce a new cultivar. Major savings are possible if single traits can be altered effectively in an otherwise outstanding cultivar. This will reduce the time required to transfer a new trait and to test a new cultivar for good adaptation if only a single trait is altered. If, however, recombinant DNA-based methods are used to produce new, but untested gene combinations, suitable testing will still be required. For major time savings to be realized, alterations at the molecular level must be predictable and precise rather than random.

Methods of Identifying Superior Individuals and Families

A most difficult task is the identification of individuals that are genetically superior, because for most traits, the true genotypic value is masked by other confounding plant traits or environmental factors (Allard & Bradshaw, 1964). One method of identifying a superior plant is to observe its performance relative to other experimental plants and to a widely grown standard cultivar. If a single observation is inconclusive, replicated observations provide a better estimate of its genetic worth. For asexually propagated plants, identical propagules can be grown for repeated observations. If asexual reproduction is not feasible, a progeny test can be used.

Recent developments in plant molecular genetics provide a promising approach for identifying genetically superior plants. The term *restriction fragment length polymorphism* (RFLP) has been coined to describe the variation among individual plants in their genomic DNA, which has been extracted, digested with restriction enzymes, blotted onto a membrane and visualized for differences in size of the DNA fragments resulting from the enzymatic cleaving at different sites on the DNA molecule (Helentjaris et al., 1986). The RFLPs have been used for mapping of plant chromosomes, establishing RFLPs as genetic markers for identifying specific stocks and cultivars, and determining linkages between RFLPs and genes or chromosome segments that control trait expression (Burr et al., 1983; Helentjaris et al., 1985). This approach appears to be particularly useful for identifying genes that control quantitative traits with low heritability (Helentjaris et al., 1986; Tanksley & Hewitt, 1988).

Breeding Methods

Pedigree breeding is used commonly for the improvement of self pollinating crops or inbred lines of cross-pollinating crops. Two parents which possess favorable, complementary traits are crossed to produce an F_1. An F_2 population is produced by selfing one or several F_1s or by intercrossing two F_1s (sib mating). Selection of the best individuals is usually begun in the F_2 population; then beginning in the F_3, the best individuals in the best families are selected. Replicated testing of families often follows in the F_4 generation to improve the effectiveness of selection for traits with low heritability. At an advanced stage of inbreeding (i.e., F_6 and F_7), the best lines or mixtures of phenotypically similar superior lines are tested for potential release as new cultivars.

Mass and recurrent selections can be used to improve populations of either self- or cross-pollinating crops. A genetically variable population of heterozygous individuals is either identified or created by intercrossing several different parents. The best plants are selected based on individual superiority, outstanding progeny, or excellent combining ability. The selected plants are intercrossed to produce a new population in which further cycles of selection are continued.

Backcross breeding has been used to transfer genes for a simply inherited, highly heritable trait into a desirable homozygous cultivar or inbred line which is the recurrent parent. The source of the trait to be transferred is called the *donor parent*. After the initial cross, individuals possessing the phenotype of the donor parent are selected and repeatedly crossed (backcrossed) to the recurrent parent. The resulting plant is expected to have the attributes of the recurrent parent (e.g., cultivar) and the desirable trait transferred from the donor parent.

Descriptions of other breeding methods that are commonly used for different traits and crops can be found in one of several reference books (e.g., Allard, 1960; Simmonds, 1979; Sneep et al., 1979; Fehr, 1987).

Testing Potential Cultivars for Commercial Utility

Proper testing should detect any major faults and establish the level of superiority or improvement over current cultivars. In addition to showing superior performance, there must be a demand for a new cultivar that is compatible with industry standards or which creates a new market. The introduction of a new cultivar will incur additional costs to the seed producer, the grower, processor and consumer; for special advertising and marketing, altered seed and commercial production practices, and new product utilization. The testing preceding release of a new cultivar should take into consideration research and development costs as well as technical superiority of the final cultivar. For seed-propagated cultivars, it must be feasible to produce seed easily and economically. Asexually propagated cultivars must be kept pest free and be able to be produced efficiently in sufficient quantities to meet demand.

CROP CULTIVARS

Definition and Types of Cultivars

Often the terms *variety* and *cultivar* are used interchangeably, but cultivar (derived from cultivated variety) is preferred for clarity, since variety is a subdivision used in botanical classifications of plant taxa. "A cultivar is a group of plants with characteristics that are distinct, uniform and stable" (Fehr, 1987, p. 377). Not only are these three properties used to define a novel variety for sexually reproduced crops covered in the Plant Varietal Protection Act (PVPA), but they also represent the attributes of successful landrace cultivars.

A cultivar is distinct in that it differs by one or more characteristics from all other cultivars. The degree of distinctness may range from subtle differences detectable only by extensive comparisons in a few environments to extreme differences that are immediately evident in many situations.

Uniformity means that any variability among plants can be described, it is predictable and for practical purposes it is acceptable commercially. Uniformity need not mean that all plants in a population (cultivar) are exactly alike genetically. It may mean that a cultivar is uniform in the sense that it is comprised of different genotypes in a known proportion that imparts an acceptable amount of variation that is unchanging over time or space.

Stability implies that when the cultivar is reproduced or reconstituted, it will remain true to the essential and distinctive features with considerable reliability commensurate with that of cultivars of the same category in which the same breeding method is employed.

Landrace Cultivars

Landrace cultivars have evolved through natural selection and the efforts of farmers to either consciously or inadvertently select for stable overall performance. This broad definition does not distinguish between types of natural reproduction; there are landrace cultivars of self- or cross-pollinated plants that have essentially the same genetic background and a high degree of homozygosity because of inbreeding. A pure line cultivar refers to a cultivar of a self-pollinated crop in which all individuals are of a single (or perhaps a few similar) homozygous genotype. Inbred line cultivars are cultivars of normally cross-pollinated crops that are highly homozygous (inbred) and of a single (or a few similar) genotype. Cultivars of apomictic crops such as Kentucky bluegrass (*Poa pratensis* L.) are considered pure line cultivars if the level of apomixis exceeds 95%.

Open-pollinated Cultivars

Open-pollinated cultivars refer to populations of normally outcrossing or mixed crossing plants that freely interpollinate to produce genetically heterozygous plants. Nevertheless, the cultivar has distinct and predictable characteristics to differentiate it from other cultivars.

Synthetic Cultivars

Synthetic cultivars may be either first or advanced generation populations, usually of cross-pollinating crops. A synthetic cultivar is produced by intercrossing a group of selected parental genotypes to produce offspring consisting of hybrids, siblings, or self-pollinating individual, depending on the degree of imposed outcrossing. The Syn-1 generation refers to the seed produced by intercrossing the parents, and the Syn-2 , Syn-3 etc. to the subsequent generations produced from open-pollination.

Hybrid Cultivars

Hybrid cultivars refer to the first filial (F_1) generation produced through controlled pollinations of select (usually inbred) parents. A single cross hybrid results from the cross between two inbred lines; a double cross hybrid from the cross between two single cross hybrid parents; and a three-way hybrid from a cross between a single cross parent and an inbred parent. Other hybrid cultivars are produced by crossing an inbred line or a single cross hybrid with an open-pollinated population, clonal cultivars, and two or more species. An F_2 cultivar is produced from the self- or sib-pollination of F_1 plants.

Multiline Cultivars

A multiline cultivar consists of a mixture of genetically and phenotypically similar (near-isogeneic) lines of a self-pollinated crop, each present in excess of 5% of the total population. A "blend" is similar to a multiline cultivar except that the components of the blend are often more diverse than those of a multiline cultivar.

Clonal Cultivars

Clonal cultivars are by definition genetically indentical and, therefore, highly uniform, except for the occurrence of somatic mutations or sports. The members of a clonal cultivar are plants that have been propagated asexually by cuttings, tubers, bulbs, grafts, rhizomes, and micropropagation.

Factors Affecting the Type of Cultivar Used Commercially

The primary factors that determine the type(s) of cultivar used for commercial production of a crop include; (i) mode of reproduction, (ii) amount of inbreeding depression and heterosis, (iii) amount and type of seed required for commercial planting, and (iv) nature of the commercial crop. For a crop species that shows obligate outcrossing, an open-pollinated cultivar will be used for a seed-propagated crop and a clonal cultivar for one propagated asexually. If the natural outcrossing can be manipulated to allow inbreeding via self- or sib-pollination, inbred lines can be produced for use as parents to produce hybrid and synthetic cultivars. For the production of hybrid cultivars, which requires that a large proportion of plants (>95%) are true hybrids (as opposed to self or sibs), pollination must be controlled either mechanically or biologically. Mechanical control of pollination is usually done through hand removal of the male inflorescence [e.g., corn (*Zea mays* L.) tassel] or anthers [e.g. tomato (*Lycopersicon esculentum* L.)]. Biological control of pollination can be through the use of ctyoplasmic-genic male sterility {e.g. corn, sorghum [*Sorghum bicolor* (L.) Moench], beet (*Beta vulgaris* L.), carrot (*Daucus carota* L.), onions (*Allium cepa* L.)}, genic male sterility (tomato), self-incompatibility mechanisms in the flower [e.g., cabbage (*Brassica oleracea* var. capitata L.) and red clover (*Trifolium pratense* L.)], and control of male and female sex expression [e.g., cucumber (*Cucumis sativus* L.)].

In autogamous or naturally self-pollinating plants [e.g., wheat (*Triticum aestivum* L.), rice (*Oryza sativa* L., bean (*Phaseolus vulgaris* L.), and oat (*Avena sativa* L.)], commercial cultivars are usually pure line or multiline cultivars. Enforced cross-pollination is practiced for experimental purposes, followed by natural self-fertilization during selection, which leads to a high level of uniformity in the final population. Sometimes it is feasible to produce commercial F_1 hybrids of self-pollinating crops if adequate seed can be produced via mechanical pollination control (e.g., tomato) or biological pollination control [e.g., cytoplasmic-genic male sterility of wheat (*Triticum* spp.)].

The production of F_1 hybrids is economically attractive for crops that express useful levels of heterosis. It is necessary to inbreed enough to produce uniform parents that can be intercrossed to produce superior F_1 hybrids. Inbred line cultivars are not used directly for commercial production of crops that show considerable inbreeding depression (e.g., corn and carrot), but for other crops where self-fertilization is the usual mode of reproduction, inbreeding depression is of little consequence.

The amount of seed required for commercial planting influences the type of cultivar used. For example, because total tomato production is on a relatively small acreage and the plants are grown at wide spacings, only a modest amount of tomato seed is required. Because a single pollination results in many seeds, it is commercially feasible to produce enough F_1 hybrid tomato seed for commercial planting using hand emasculation and pollination. In contrast, wheat is grown on large areas of land and sown at high densities. Using mechanical pollinations that result in one seed per cross, it is not possible to produce nearly enough hybrid seed for commercial planting. Therefore, the feasibility of F_1 hybrid seed wheat production rests on the use of biological pollination control mechanisms such as chemical emasculation or cytoplasmic-genic male sterility.

The nature of the commercial product derived from each crop also influences the type of cultivar used. Sometimes, a high degree of uniformity is required, such as with fruits and vegetables or crops used for special purposes [e.g., barley (*Hordeum vulgare* L.) for malting]. Uniformity can be expected in pure line cultivars of self-pollinating crops [e.g., lettuce (*Lactuca sativa* L.) and bean], in F_1 hybrid cultivars of cross-pollinating crops (e.g., onion and carrot), and in clonal cultivars [e.g., fruit trees and potato (*Solanum tuberosum* L.)]. Where pollination control is not practiced, genotypic and phenotypic variability result.

Some genetic variability within and/or among populations is desirable. Where there is a threat to production by biological or abiotic agents such as pests or unpredictable weather, genetic uniformity may produce undesirable genetic vulnerability (NAS, 1972).

Reproducibility and Predictability

Successful cultivars persist in the marketplace not only because of superior productivity, but also because their performance is reproducible and predictable. An important factor in the sustained production of new field crop cultivars has been the presence of seed certification programs in the USA and several other countries. Those associations were established to assure authenticity of cultivars, trueness to type, and seed purity. As commercial companies have expanded into seed sales, they have assumed those responsibilities because a good reputation is essential for their continued vitality.

Superiority and Uniqueness

Uniqueness and superiority help a new cultivar gain a share of the market. Degree of uniqueness is important, since a new cultivar must be different enough to distinguish it from the competition, yet similar enough to fit the standards of an established market. Sometimes a new cultivar is unique enough to produce a new product on which a totally new market is established.

For some crops where high productivity and quality prevail, it is difficult to develop outstanding new cultivars that are greatly superior. Then, slight improvements for other traits that provide better productivity in an important agroecological area or stable production over a wider geographic area are important. Large increases in performance often await a major breakthrough in breeding or crop management.

Proprietary Control of Cultivars

Control and protection of cultivars may be sought either to maintain seed purity and identity, or to obtain remuneration to the developer for the commercial product. Often the control of a new cultivar by the breeder effectively ends when it is released for commercial use. Then it is essential that a suitable organization increase stock seeds and produce commercial quantities of seed for marketing and distribution to growers.

Proprietary control of cultivars can be by either biological or statutory means. Biological control is through controlled access to parents for production of F_1 hybrid seeds (e.g., corn, onion, and cabbage); exclusive use of mechanisms for pollination control (e.g., the systems of male sterility), and the production and sale of seeds that can be used for production of only one crop (e.g., F_1 hybrids). The production and sale of F_1 hybrids provides an effective means of proprietary control since there are few, if any, opportunities to circumvent the system.

The Plant Patent Act of 1930 covering asexually propagated plant species, the 1970 PVPA for sexually propagated plants, and the basic patent act (Sec. 101) offer legal means for protecting plant cultivars. These laws are enforceable to the extent that the holder of proprietary rights wished to do so. Although it may not be economically feasible to pursue every infringement, the loss can be substantial if a large number of violations occur. In contrast, with biological control even the small user must pay the added cost to grow a superior F_1 hybrid cultivar.

Distribution of Cultivars

The distribution of cultivars usually follows one of several routes. The price of seed to the buyer will reflect at least the cost of seed production. The responsibility for production is sometimes borne by a public agency, such as a foundation seed organization or a governmental agency. In that case, seed cost will reflect a nonprofit situation, with prior costs of research and development being paid through taxpayer support.

For crop cultivars distributed through private companies, seed costs include the cost of production, advertising and marketing, and profit to the grower and distributor. The price reflects competition and cultivar uniqueness, superiority, reproducibility, and predictability. In addition, there is a cost for research and testing required to develop the cultivar.

For some crops, seed sales among farmers are prevalent, e.g., winter wheat and oat. For those crops (e.g., lettuce and snap bean) in which seedling quality is crucial and can be assured only if seeds are produced by well-trained technicians, specialized companies are the primary distributors. Some companies conduct basic and applied research, while others use public breeding lines released by government agencies for production of commercial seed. Other companies buy seeds wholesale, then treat and repackage them for sale to growers (e.g., the home gardener).

REQUIREMENTS FOR SPECIAL CROP PLANTS

The procedures discussed heretofore apply to the breeding of most crop plants, but the specialized needs of some crops must be recognized. A sizeable number are categorized as minor crops because they do not contribute to large food or industrial uses on a worldwide scale. Nevertheless, they are important for a particular group of people or a specialized use.

Limited Demand, Small Acreage, and Narrow Adaptation

Some crops have a specialized use, are adapted to a narrow agroecological area or are grown on a limited acreage. These include field crops such as hops, many of the grain legumes, safflower, etc.; many of the vegetables and most of the ornamental plants and flowers. Crops such as red beet, artichoke (*Cynara scolymus* L.), okra [*Abelmoschus esculentus* (L.) Moench]; and even the most important vegetables including tomato, sweet corn, onion, bean, pea, etc., are "minor" based on acreage and total value compared to the large field crops.

Crops with Low Direct Value

Some crops are important but because of their low direct value, it may be difficult to justify research and development costs without understanding how they are used. Forage grasses and forage legumes have limited direct value based on the sale of hay or silage, yet they are essential to the beef (*Bos taurus*), dairy (*Bos taurus*), and sheep (*Ovis aries*) industries. They must be evaluated in light of their contribution to the animal industry. Other examples are turfgrass and shade trees that are valued for aesthetic purposes in yards, greenbelts, parks, and golf courses.

Crops with Long Life Cycles

Extended life cycles and long periods of juvenility hinder the potential for improving many of the fruit, nut, and forest tree crops through conventional breeding procedures. The time required for cultivar development using conventional breeding is much longer than for annual and biennial plants. Not only is the breeding cycle extended by several years, but also the time for testing is longer. Instead of a 7 to 10 yr time-frame needed to produce a new field crop or vegetable cultivar, it may take 10 to 15 yr to produce a new tree crop if all goes well. Because of the size of the trees, more land is required for research, which coupled with the time factor produces a greater cost of cultivar development of woody plants. The opportunities to use recombinant DNA-based methods to reduce time for production and testing of new cultivars is especially attractive in these crops.

Limited Potential for Commercialization

Some crops and certain desirable crop traits have limited potential for commercialization, but nevertheless, are valuable to society. For example, carrots are a primary source of β-carotene, but because they are a minor crop and vitamin A can be supplemented in diets of relatively affluent people, development of high-carotene carrots receives little attention. Nevertheless, in low-income countries blindness associated with insufficient vitamin A is often pervasive. For many people, consumption of high-carotene carrots would be both cost effective and an attractive solution to an important medical problem.

Special crop plants are mentioned here to emphasize that special treatment may be justified. An exclusive or semi-exclusive release or licensing agreement may provide the incentive for a company to market a crop cultivar having limited commercial value. Without assurance of a reasonable return to a single company, a new cultivar of a minor crop may never reach the marketplace.

GENETIC FOUNDATIONS OF CROP IMPROVEMENT

Genes and Chromosomes

Plant breeding can be looked at as the alteration of the genetic organization and composition of crop plants. Mendelian genetic principles provide the basis for modern plant breeding. Heritable traits are transmitted from parent to offspring, where they are expressed in predictable ratios. Segregation results from reassortment of genes or heritable units at meiosis during sexual reproduction. Alleles are the alternative forms of a gene. The traits displayed in the offspring arise from various combinations of dominant alleles (which express when only one copy is present) and recessive alleles (two copies are required for expression). Two or more traits may show independent assortment when transmitted to the offspring; but when both genes are located in close proximity on the same chromosome, a preponderance of parental types results from genetic linkage. Cytogenetics combines cellular and subcellular studies (cytology) with genetic analyses. The chromosomes, which are located in the nucleus provide the physical basis for heredity.

The chromosome theory of heredity, which was established between 1910 and 1915, provides that: (i) genes controlling distinct traits show the same patterns of transmission and segregation as specific chromosomes which are identifiable by distinctive physical features, and (ii) rearrangements (recombination) occur between linked genes as a result of physical exchange (crossing over) between homologous chromosomes. In 1944, deoxyribonucleic acid

(DNA) was first reported to carry genetic information, and in 1953, the structure of DNA was revealed (see Watson et al., 1983). The demonstration that DNA forms a double helix held together by hydrogen bonds between specific base pairs, provided the model necessary to explain the template for production of proteins.

Non-Mendelian Inheritance

Some heritable traits are controlled neither totally nor in part by nuclear genes. Rather, the offspring resemble the maternal (seed) parent and these traits are controlled by DNA contained in the chloroplasts and mitochondria located in the cell cytoplasm. For traits controlled completely by cytoplasmic factors, the offspring resemble the female parent for many subsequent generations unless mutations occur or nuclear genes are introduced which modify the cytoplasmic control. Examples are cytoplasmic male sterility of several crops (e.g., corn, sorghum, and beet), leaf chlorophyll patterns and reaction to certain plant diseases [e.g., southern corn leaf blight (*Helminthosporium maydis*) of corn].

Sometimes the first generation offspring (F_1) are influenced by the maternal genotype or the physiological state of the seed parent, but this maternal effect is not transmitted to the next generation. On some occasions, there is an effect of the male or pollen parent on the seed or fruit that is called *xenia* and *metaxenia,* respectively.

Molecular Genetics

Molecular genetics deals with studies at the molecular level beginning with nuclear, mitochondrial, and chloroplast DNA, encompassing the intermediate stages of gene regulation and expression, and culminating with production of gene products that account for the final plant phenotype. Included in these studies are: (i) structure and function of the nuclear and mitochondrial genomes, (ii) structure and expression of nuclear genes, (iii) chromatin structure, and (iv) gene regulation. An understanding of the molecular basis of developmental biology will help elucidate how phenotypic diversity in form and function can arise from expression of the genetic code.

Quantitative Genetics

Quantitative genetics, or the genetic control of quantitative trait expression, is commonly associated with the use of statistical or biometrical methods to analyze the continuous variation displayed in segregating populations. There is a tendency to distinguish sharply between traits that show continuous variability and those that show discrete segregation, hence quantitative vs. qualitative genetics. Although the biometrical approach to analysis of quantitative variation is useful for describing population characteristics, drawing inferences based on relationships among individuals or families, and for making predictions about future generations, this approach gives no indication of the biological process (Wright, 1968).

The multiple gene hypothesis has been used to explain the basis for quantitative expression of traits, and, consequently polygenic control and quantitative expression are often inextricably linked. Although quantitative traits are usually found to be less than completely heritable, indicating that nongenetic factors as well as genetic segregation influence trait expression, multiple genes need not be segregating to produce continuous variation if there is a large contribution of nongenetic variability because of environmental fluctuations or other factors. Other assumptions in addition to the multiple factor hypothesis are that the effect of each gene is small, all gene effects are equal, and they are cumulative to produce the total genotypic effect. While these assumptions are convenient for constructing a biometrical model, they provide limited opportunity to study the effects of individual genes and to manipulate genes at the molecular or cellular level.

Biometrical approaches to quantitative inheritance are powerful tools for understanding the response to selection imposed in natural and artificial populations. The ability to identify parents that produce superior hybrid offspring (combining ability) has been important for using heterosis or hybrid vigor in the production of high-performing hybrid cultivars in crops such as corn, sorghum, beet, and onion.

COMPONENTS OF VARIABILITY

The phenotype is the observed or measured attribute of the plant trait or combination of traits; and the variation for those attributes among individuals or families of plants is phenotypic variation. Most traits are complex, because each results from interaction of several related traits that contribute to the final expression. For example, grain yield of a cereal plant is the result of three principal components, number of tillers (stems) per plant, number of seeds per tiller, and size or weight of each seed, and each of those principal components is itself complex. The level of complexity depends on the different types of gene expression, the number and type of metabolic pathways that control the trait and the extent to which environment affects the plant. Each trait and component may show different patterns of inheritance depending on genetic segregation and the amount of influence by nongenetic factors.

Environmental Variation

The heritable part of phenotypic variation is contained in the genetic code and the remainder is because of the environment in which the plant is grown (Bell, 1977). The expression of some traits is affected little by the environment (highly heritable traits), while others are affected greatly (traits with low heritability). For the latter, it is difficult to determine the exact proportion of the phenotypic variation due to genotype. When all the phenotypic variability is because of nongenetic causes, no improvement from selection can be made because there is no genetic variability or heritable differences among plants. Such is the case in populations of plants that are asexual propagules from a single parent plant, e.g., potato plants from one tuber, grafted fruit trees.

When environmental effects obscure the genetic value, it is helpful to make repeated observations on the plant growing under different conditions (i.e., replications in time and space). Multiple year and location observations provides a better estimate of the genetic value of each plant because the potential of the genotype can be observed under both favorable and unfavorable environments. Another way to better estimate the genotypic value of an individual is to observe the phenotypes of its progeny. Better estimates result when more progeny are evaluated.

Genetic Variation

Continuous vs. Discontinuous Variation

Populations that are genetically variable for a trait can display continuous or discontinuous variation. When variation is discontinuous, individuals or families can be assigned to discrete classes among which there is no overlap between class boundaries. With continuous variation, there are no obvious groupings regardless of the precision of measurement of the trait expression. Continuous variation will often form a normal distribution that may be skewed toward either extreme.

Qualitative vs. Quantitative Inheritance

As noted previously, traits are often considered to be either qualitative or quantitative. For qualitative traits, there is usually discrete segregation in the F_2, F_3, and backcross generations. The segregation patterns can be accounted for by one or a few genes, with alternative alleles showing dominance, recessiveness and sometimes interactions between different loci (i.e., epistasis). Qualitative traits are usually highly heritable, approaching 100%. Large phenotypic changes for traits such as composition of the embryo and endosperm of seeds (e.g., Nelson, 1969; Brown et al., 1982; Boyer & Shannon, 1983; Wilcox & Cavins, 1985), composition of roots and tubers (e.g., Freeman & Simon, 1983), and appearance and ripening of a certain fruits (e.g., Tigchelaar et al., 1978) can result from segregation at a single locus.

Quantitative inheritance is characterized by a continuous distribution of individuals in a segregating population, and is a consequence of either a large number of segregating loci controlling the trait, a large influence of nongenic factors, or a combination of both. Even though individual gene effects may be small relative to the total variation, transgressive segregation (the occurrence of genotypes more extreme than either parent) coupled with selection, is a powerful force for altering trait expression in a population. The continuing response to long-term selection for high and low percentage seed oil and protein in a corn population, beginning in 1896 and continuing until present, has resulted in plants with greatly different seed composition (Dudley, 1977). Heritability of quantitative traits is always $<100\%$ and often is low, approaching zero in some cases.

Components of Quantitative Genetic Variance

The total genetic variance can be divided into components (Dudley & Moll, 1969). Additive genetic variance results from the linear effect of substituting one allele for another at a specific locus, with the value of the heterozygote being intermediate between the two homozygotes (i.e., $aa = 0, Aa = 1, AA = 2$). Dominance variance results from deviations from linearity or additivity of allelic substitutions at a single locus (i.e., $aa = 0, AA = 2, Aa \neq 1$). A third component of genetic variance, epistatic variance, results from interaction of segregating alleles at two or more different loci to produce a combined effect that is different from the effects of each locus considered separately.

SOURCES OF GENETIC VARIATION

Naturally Occurring Variation

For most crop plants, there is extensive naturally occurring genetic variation. This variation results from gene mutation followed by genetic recombination and from somatic mutation. There are often strikingly similar characters found in related species. This "law of homologous series in variation" that was stated in genetic terms by Vavilov (Allard, 1960, p. 26) is useful in predicting that certain characters found in one species might be expected in a similar species. For example, the endosperm mutants that affect seed protein composition of corn (i.e., opaque-2 and others) (Nelson, 1969) have counterparts in other cereals (e.g., the Hiproly in barley and opaque mutants in sorghum (Munck et al., 1971; Singh & Axtell, 1973). Natural selection enhanced by the mating system(s) characteristic of plant communities, provides the forces that generate and maintain genetic variability.

Mating systems, dependent on the type of reproduction, influence the amount and type of variation in plant populations. Plants of some species propagate asexually through the division of vegetative parts, e.g., bulbs, roots, stems, to produce communities that are genetically homogeneous except for occasional sports or somatic mutants. Some species such as Kentucky bluegrass reproduce primarily through apomixis, a process in which seeds are produced from maternal cells. Again, all offspring (seeds) will be uniform and like the maternal parent. Plants that normally reproduce sexually are either predominantly self-, cross- (out-crossing), or mixed-fertilizing (both self- and cross-fertilizing). Plants of a self-fertilizing species tend to be highly homozygous (uniform), and populations contain mixtures of near-homozygous plants. Plants of a cross-fertilizing species are heterozygous and the populations heterogeneous. Heterozygosity provides the opportunity for a recessive allele to be masked by a dominant allele, and upon self-fertilization, the recessive alleles are expressed when homozygous. Cross-fertilizing species often show considerable inbreeding depression or reduced vigor upon inbreeding (mating of genetically related individuals), due largely to the uncovering of deleterious recessive alleles. There is usually little inbreeding depression in self-fertilizing species since deleterious recessive alleles cannot escape detection by being masked.

Heterosis is the superiority of the hybrid offspring compared to the parents or the population from which the parents were derived. Usually heterosis is most pronounced in naturally cross-fertilizing species, but not exclusively so. The amount of heterosis depends on the genotypes of the parents, with the ability of a plant to produce superior progeny when combined with a wide array of other plants being called good general combining ability (GCA). Plants that produce superior offspring when crossed with one or a few other genotypes are said to have high specific combining ability (SCA). Although heterosis has been known for many years, and it is the foundation upon which some major seed industries are built (e.g., hybrid corn, sorghum, and beet), the complete biological basis for heterosis is still not known. Combining ability is a heritable trait that can be improved through selection, but it is a complex trait with low to medium heritability. Often combining ability is poorly related to parental performance per se, and progeny tests are required to identify plants that have good combining ability.

Domesticated Populations

Domesticated populations are an important source of genetic variability. Landrace cultivars are groups of related plants that are genetically different from other such groups and which have arisen as a consequence of selection by farmers for local geographical adaptation. Other types of domesticated populations include breeder produced cultivars that are currently popular and older cultivars that have been displaced by more recent superior ones. The breeding populations that are maintained by plant researchers are another source of genetic variability that is ready for rapid incorporation into potential new cultivars.

Nondomesticated Forms and Related Taxa

Wild and nondomesticated relatives of cultivated materials are usually sexually compatible with cultivars, but are not usable directly because of numerous "wild" traits that make them undesirable for cultivation. Effective gene transfer is usually more difficult and time consuming than from adapted cultivars because of the need to exclude undesirable genes during transfer via sexual hybridization. Related taxa, either cultivated or nondomesticated materials, are other sources of variability. The difficulty in transferring genes depends on factors such as adaptation to the environment, degree of genetic homology, synchrony of flowering, type of pollination, likelihood of fertilization, pollen compatibility, embryo-endosperm growth, regularity of meiosis, chromosome pairing, and hybrid viability.

Wide Hybridization

Wide hybridization refers to crosses between distantly related taxa that are not easily accomplished through normal crossing and sexual hybridization (i.e., interspecific, intergeneric, and interfamilial crosses). In some cases, pollen will not germinate on the style of an unrelated female plant, or pollen tube growth may be impaired prior to fertilization. Even after fertilization and initiation of cell division, the endosperm tissue that serves a nourishing role to the embryo may cease to develop, with embryo abortion occurring soon thereafter. After initiation of embryo growth, it may be possible to excise the embryo prior to complete disintegration of the endosperm and to culture the embryo on artificial media to produce a viable hybrid plant. Embryo rescue has been used to produce hybrids between early and late maturing stone fruits (Turkey, 1933) and for the initial crosses between wheat and rye to produce the new crop known as triticale (× *Triticosecale rimpaui* Wittm.) (Larter & Gustafson, 1980). Extensive use of sterile culture and artificial media has broadened the use of embryo rescue to facilitate wide hybridization.

The different forms of naturally occurring variation described above are maintained in in situ collections and germplasm banks in various countries throughout the world. These collections have been an invaluable source of genes for improving a wide range of traits in many crop plants. See Fehr (1987) for a description of the U.S. germplasm system.

Enhanced Genetic Variation

Induced Mutation

Fehr (1987) states that the average rate of mutation for an individual gene is approximately 10^{-6}. Although natural mutation results in new variability, these frequencies are too low for routine use in plant breeding during a limited time scale. The discovery that certain chemical and physical agents are mutagenic or produce heritable changes led to the use of mutagens to increase the frequency of mutation in crop plants. Different mutagens induce different types of mutation (e.g., ionizing radiation produces chromosome damage and chemicals such as ethyl methane sulfonate [EMS] produce point mutations) (Manual on Mutation Breeding, 1977), but there is little evidence that induced mutations are unique.

The production of induced mutations for plant improvement has been limited primarily to asexually propagated and self-fertilizing plants. Because the frequency even of induced mutation is usually of the order of 10^{-3} or less, large populations must be mutagenized and evaluated, and the mutant phenotypes must be easily detectable for this approach to be effective. Therefore, induced mutations have been most useful for improving traits that can be identified visually or with a rapid screening method, such as plant stature, leaf shape and color, chlorophyll reduction, and changes in seeds and fruits.

Although the use of induced mutations was proclaimed as a major advancement for planting breeding, and apart from the fact that some cultivars of asexually propagated and self-pollinating plants have been produced, mutation breeding has not made a major contribution to crop improvement. Nevertheless, there are specific situations where the use of induced mutations has been important, i.e., the transfer of disease resistance to cultivated wheat on a small chromosome segment from the wild species *Aegilops umbelluata* (Sears, 1956).

Somaclonal Variation

Plant cell and tissue culture have become important methods used in plant research. Growth of cell or tissue explants on an artificial medium may be viewed as a special type of asexual propagation, and, therefore, the expectations are that derived plants should remain true to type. This is not always the case and sometimes regenerated plants are quite different from the progenitor (Smeal, 1986). Larkin and Scowcroft (1981) proposed the term ". . . somaclones for plants derived from any form of cell culture. Somaclonal variation is then the variation displayed amongst such plants." Although somaclonal variants are somewhat similar to spontaneous and induced mutations in vivo, the frequency of occurrence is usually higher. In most cultures, the media contain no known mutagenic agents. By definition, the occurrence of somaclonal variants is limited to the crop plants amenable to plant regeneration in relatively large numbers such as tobacco (*Nicotiana tabacum* L.), sugarcane (*Saccharum officinarum* L.), potato, and barley.

There is no apparent single underlying basis for somaclonal variation. The origins of somaclonal variation, as discussed by Larkin and Scowcroft (1981), include karyotype changes such as aneuploidy and polyploidy, cryptic-changes in the chromosomes, action of transposable elements, somatic gene rearrangements, gene amplification and deletion, and cryptic virus elimination. Major reservations about the practical use of somaclonal variation include its random occurrence and instability. If a crop is produced commercially through asexual means, the changes need not be transmissible through meiosis, as is required for seed-produced crops. Although the changes that occur may produce increased variation, many variants are of little or no practical value. In some cases, selection can be exerted in populations where somaclonal variation occurs. Because many plantlets can be grown in a small space and under defined conditions, selection for traits such as resistance to pathogen toxins and tolerance to stress-inducing factors, i.e., salt, heavy meatals may be effective where unique useful variation occurs. Overall, the extent to which somaclonal variation will be used for crop improvement is not evident.

Genetic Transfer by Nonsexual Methods

The ability to culture plant cells, tissues and organs in vitro has contributed greatly to the development of methods for the nonsexual transfer of genes between plants. Two general systems of genetic transfer using nonsexual methods are cell fusion and introduction of foreign DNA directly (Goodman et al., 1987). With cell fusion, it is necessary to prepare large numbers of single protoplasts (i.e., cells stripped of their cell walls). The protoplasts are induced to fuse using certain chemicals or electrical current. To be utilized further, a whole plant must be regenerated from the fusion product. Although considerable work has been done on cell fusion, there are few examples of the practical application to crop improvement. One potential use would be the combination of the nuclear genome of one taxon with the cytoplasm of another to produce cytoplasmic male sterility. Another possible use is the combination of nuclear genomes to produce a plant with an increased ploidy level.

Several problems are associated with the practical use of cell fusion. It is difficult to regenerate plantlets of many species or the fusion products of widely different taxa. The method is imprecise and involves the transfer of the whole genome including deleterious as well as favorable alleles. Unless fusion is between haploids, the resulting product will have a higher ploidy level than either parent. Often sexual reproduction is difficult in the plants resulting from wide combinations.

The second approach to gene transfer using nonsexual methods involves introduction of known DNA sequences directly. The important features of nonsexual DNA transfer techniques are that: (i) it extends the range of genetic variability beyond that normally available through sexualy hybridization, and (ii) there is a high degree of precision possible when only the DNA comprising a single gene is transferred. A reliable means of DNA transfer is of absolute necessity. Several methods are used including: Agrobacterium-mediated gene transfer, virus-mediated gene transfer, DNA transfer where purified DNA is introduced into protoplasts through direct uptake from the surrounding medium, electroporation or microinjection (Uchimiya et al., 1986; De la Pena et al., 1987; Hinchee et al., 1988; Goodman et al., 1987), and the use of high velocity microprojectiles for delivering nucleic acids into living cells (Klein et al., 1987, McCabe et al., 1988). Most methods are not generally applicable to many different plants, and the inability to routinely transfer specific DNA to many economically important plants is still a hindrance to manipulating DNA directly.

Another difficulty in using direct DNA transfer is the paucity of well-characterized plant genes available for transfer. Also, genes and/or gene products are difficult to detect, thus preventing the easy recognition of transformed cells or tissues. Selectable traits such as resistance to certain substances that are added to the culture media (e.g., kanamycin, methotrexate, and hygromycin) (Reiss et al., 1984; Morrish et al., 1987), expression of the beta-glucuronidase (GUS) gene (Jefferson, 1987) and others, are used to facilitate detecting transformants. Genes for the selectable traits are ligated to the genes of interest in the transformation vector. By selecting for transformants showing the selectable marker trait, the likelihood of obtaining transformants having the gene of interest is increased. These candidates are subjected to more rigorous molecular analysis for further confirmation.

The use of direct DNA transfer methods that involve transformation of single cells has been limited to crops in which plant regeneration can be done easily (e.g., solanaceous crops). Since regeneration is difficult in some of the important crops, e.g. corn, wheat, soybean [*Glycine max* (L.) Merr.], methods for introduction of DNA into tissues capable of continuing differentiated growth have received more interest (McGranahan et al., 1988; Ulian et al., 1988; McCabe et al., 1988; Hinchee et al., 1988).

Although there are no examples of a commercial crop cultivar developed through genetic transfer using direct manipulation of DNA, several experimental systems are promising. Bacterial genes for insecticidal protein from *Bacillus thuringiensis* have been engineered into several plants (Barton et al., 1987; Umbeck et al., 1987). Tobacco and tomato plants have been engineered for virus resistance using the gene encoding the coat protein, antisense RNA and satellite RNA (Nelson et al., 1988; Cuozzo et al., 1988; Gerlach et al., 1987). Genes conferring resistance to chemical herbicides have been engineered into crop plants (Fillatti et al., 1987). How rapidly plants produced by transformation will gain acceptance in the marketplace depends on their relative superiority, the time needed to evaluate them under field conditions and the skill in distributing and marketing new cultivars to growers and consumers.

Transposable Elements

Although chromosomes are usually regarded as being stable structures in which the genetic information is precise, it is now known that DNA components can undergo specific types of modification and rearrangement during development or in response to stress (Nevers & Saedler, 1977; McClintock, 1978). Based on corn studies, McClintock postulated that transposable controlling elements were responsible for a series of changes in gene expression. A striking characteristic of the controlling elements is their ability to move about in the genome. These elements have been shown to alter the DNA organization within loci to produce new alleles and gene products and to enhance the occurrence of chromosomal rearrangements, which may in turn lead to new nucleotide sequences and even new genes (Nevers & Saedler, 1977). In addition to representing a possible means for altering genetic information, some chromosomal rearrangements may play a part in cell differentiation. The nucleotide sequencing and ability to recognize transposable elements in the genome offer a means of identifying specific loci affecting a trait first by observing a change in phenotype, then detecting where the element is inserted in the DNA. Transposon tagging provides a means of detecting a gene without first knowing the gene product, a difficulty that has hampered extensive identification of many genes.

The extent to which transposable elements occur in plants is difficult to determine precisely, but their presence in several different organisms (e.g., Drosophila, corn, and nematodes) suggests that they may be widely distributed. If not, the transfer of controlling elements via transformation may offer a means of introduction into plants where they do not occur already. Although additional stable genetic variability can result from the action of transposable elements, their practical use in crop improvement is only being contemplated.

REFERENCES

Allard, R.W. 1960. Principles of plant breeding. John Wiley & Sons, New York.

Allard, R.W., and A.D. Bradshaw. 1964. Implications of genotype-environmental interactions in applied plant breeding. Crop Sci. 4:503–508.

Barton, K.A., H.R. Whiteley, and N-S. Yang. 1987. *Bacillus thuringiensus* -endotoxin expressed in transgenic *Nicotiana tabacum* provides resistance to Lepidopteran insects. Plant Physiol. 85:1103–1109.

Bell, E.A. 1977. Heritability in retrospect. J. Hered. 68:297–300.

Boyer, C.D., and J.C. Shannon. 1983. The use of endosperm genes for sweet corn improvement. p. 139–161. *In* J. Janick (ed.) Plant breeding reviews. Vol. 1. AVI Publ. Co., Westport, CT.

Brown, J.W.S., D.R. Ersland, and T.C. Hall. 1982. Molecular aspects of storage protein synthesis during seed development. p. 3–42. *In* A.A. Khan (ed.) The physiology and biochemistry of seed development, dormancy and germination. Elsevier Press, New York.

Burr, B., E. Evola, F.A. Burr, and J.S. Beckman. 1983. The application of restriction fragment length polymorphism to plant breeding. p. 45–59. *In* J.K. Setlow and A. Hollaender (ed.) Genetic engineering principles and methods. Plenum Press, New York.

Cuozzo, M., K.M. O'Connell, W. Kaniewski, R-X. Fang, N-H. Chua, and N.E. Tumer. 1988. Viral protection in transgenic tobacco plants expressing the cucumber mosaic virus coat protein or its antisence RNA. Biotechnology 6:549–557.

De la Pena, A., H. Lott, and J. Schell. 1987. Transgenic rye plants obtained by injecting DNA into young floral tillers. Nature (London) 325:274–276.

Dudley, J.R. 1977. 76 generations of selection for oil and protein percentage in maize. p. 459–473. *In* E. Pollak et al. (ed.) Proc. of the Int. Conf. on Quant. Genet. Iowa State Univ. Press, Ames.

Dudley, J.W., and R.H. Moll. 1969. Interpretation and use of estimates of heritability and genetic variances in plant breeding. Crop Sci. 9:257–262.

Fehr, W.R. 1987. Principles of cultivar development. Vol. 1. Theory and technique. Macmillan Publ. Co., New York.

Fehr, W.R., and H.H. Hadley (ed.) 1980. Hybridization of crop plants. ASA and CSSA, Madison, WI.

Fillattti, J.J., J. Kiser, R. Rose, and L. Comai. 1987. Efficient transfer of glyphosate tolerance gene into tomato using a binary *Agrobacterium tumefaciens* vector. Biotechnology 5:726–730.

Freeman, R.E., and P.W. Simon. 1983. Evidence for simple genetic control of sugar type in carrot (*Daucus carota* L.). J. Am. Soc. Hortic. Sci. 108:50–54.

Gerlach, W.L., D. Llewellyn, and J. Haseloff. 1987. Construction of a plant disease resistance gene from the satellite RNA of tobacco ringspot virus. Nature (London) 328:802–805.

Goodman, R.M., H. Hauptli, A. Crossway, and V.C. Knauf. 1987. Gene transfer in crop improvement. Science 236:48–54.

Helentjaris, T., G. Kind, M. Slocum, C. Siedenstrang, and S. Wegman. 1985. Restriction fragment polymorphisms and probes for plant diversity and their development as tools for applied plant breeding. Plant Mol. Biol. 5:109–118.

Helentjaris, T., M. Slocum, S. Wright, A. Schaefer, and J. Nienhuis. 1986. Construction of genetic linkage maps in maize and tomato using restriction fragment length polymorphisms. Theor. Appl. Genet. 72:761–769.

Hinchee, M.A.W., D.V. Conner-Ward, C.A. Newell, R.E. McDonnell, S.J. Sato, C.S. Gasser, D.A. Fischoff, D.B. Re, R.T. Fraley, and R.B. Horsch. 1988. Production of transgenic soybean plants using *Agrobacterium*-mediated DNA transfer. Biotechnology 6:915–922.

Klein, T.M., E.D. World, R. Wu, and J.C. Sanford. 1987. High velocity microprojectiles for delivering nucleic acids into living cells. Nature (London) 327:70–73.

Larkin, P.J., and W.R. Scowcroft. 1981. Somaclonal variation- a novel source of variability from cell cultures for plant improvement. Theor. Appl. Genet. 60:197–214.

Larter, E.N., and J.P. Gustafson. 1980. Triticale p. 681–694. *In* W.R. Fehr and H.H. Hadley (ed.) Hybridization of crop plants. ASA and CSSA, Madison, WI.

Manual on Mutation Breeding. 1977. Technical Report Series N. 119. International Atomic Energy Agency, Vienna.

McCabe, D.E., W.F. Swain, B.J. Martinell, and P. Christou. 1988. Stable transformation of soybean (*Glycine max* L.) by particle acceleration. Biotechnology 6:923–926.

McClintock, B. 1978. Mechanisms that rapidly reorganize the genome. Proc. 10th Stadler Symp., Univ. of Missouri Press, Columbia.

McGrahahan, G.H., C.A. Leslie, S.L. Uratsu, L.A. Martin, and A.M. Dandekar. 1988. Agrobacterium-mediated transformation of Walnut somatic embryos and regeneration of transgenic plants. Biotechnology 6:800–804.

Morrish, F., V. Vasil, and I.K. Vasil. 1987. Developmental morphogenesis and genetic manipulation in tissue and cell cultures of Gramineae. p. 431–500. *In* J.G. Scandalois (ed.) Adv. Genet. 24. Academic Press, New York.

Munck, L., K.E. Karlsson, and A. Hagberg. 1971. Selection and characterization of a high protein lysine variety from the world barley collection. p. 544–548. *In* R. Nilan (ed.) Barley genetics II. Washington State Univ. Press, Pullman, WA.

National Academy of Science. 1972. Genetic vulnerability of major crops. NAS, Washington, DC.

Nelson, O.E. 1969. Genetic modification of protein quality in plants. Adv. Agron. 21:171–194.

Nelson, R.S., S.M. McCormick, X. Delannay, P. Dube, J. Layton, E.J. Anderson, M. Kaniewska, R.K. Proksch, R.B. Horsch, S.G. Rogers, R.T. Fraley, and R.N. Beachy. 1988. Virus tolerance, plant growth and field performance of transgenic tomato plants expressing coat protein from tobacco mosaic virus. Biotechnology 6:403–409.

Nevers, P., and H. Saedler. 1977. Transposable genetic elements as agents of gene instability and chromosomal rearrangements. Nature (London) 268:109–115.

Reiss, B., R. Sprengel, H. Will, and H. Schaller. 1984. A new sensitive method for qualitative and quantitative assay of neomycin phosphotransferase in crude extracts. Gene 30:211–218.

Rennie, B.D., J. Zelka, M.M. Cramer, and W.D. Beversdorf. 1988. Genetic analysis of low linolenic acid levels in the soybean line PI361088B. Crop Sci. 28:655–657.

Sears, E.R. 1956. The transfer of leaf rust-resistance from *Aegilops umbellulata* to wheat. Brookhaven Symp. Biol. 9:1–21.

Simmonds, N.W. 1979. Principles of crop improvement. Longman Group Ltd., New York.

Singh, R., and J.D. Axtell. 1973. High lysine mutant gene (hl) that improves protein quality and biological value of grain sorghum. Crop Sci. 13:535–539.

Smeal, J. 1986. Tissue culture and crop improvement: Scientific dream of new Science? p. 8–13. *In* J. Smeal (ed.) Somaclonal variations and crop improvement. Martinus Nijhoff Publ., Hingham, MA.

Sneep, J., A.J.T. Hendriksen, and O. Holbek. (ed.). 1979. Plant breeding perspectives. Centre for Agricultural Publishing and Documentation. Wageningen, Netherlands.

Tanksley, S.D., and J. Hewitt. 1988. Use of molecular markers in breeding for soluble solids content in tomato- a re-examination. Theor. Appl. Genet. 75:811–823.

Tigchelaar, E.C., W.B. McGlasson, and R.W. Breuscher. 1978. Genetic regulation of tomato fruit ripening. HortScience 13:508–513.

Tukey, H.B. 1933. Artificial culture of sweet cherry embryos. J. Hered. 24:7.

Uchimiya, H., T. Fushimi, H. Hashimoto, H. Harada, K. Syono, and Y. Sugawara. 1986. Expression of a foreign gene in callus derived from DNA-treated protoplasts of rice (*Oryza sativa* L.). Mol. Gen. Genet. 204:204–207.

Ulian, E.C., R.H. Smith, J.H. Gould, and T.D. McKnight. 1988. Transformation of plants via the shoot apex. In vitro Cell. Devel. Biol. 24:951–954.

Umbeck, P., G. Johnson, K. Barton, and W. Swain. 1987. Genetically transformed cotton (*Gossypium hirsutum* L.) plants. Biotechnology 5:263–366.

Watson, J.D., J. Tooze, and D.T. Kurtz. 1983. Recombinant DNA. A Short course. W.H. Freeman and Co., New York.

Wilcox, J.R., and J.F. Cavins. 1985. Inheritance of low linolenic acid content of the seed oil of a mutant of *Glycine max.* Theor. Suppl. Genet. 71:74–78.

Wright, S. 1968. Evolution and the genetics of population. Vol. 1. Genetics and biometric foundations. Univ. of Chicago Press, Chicago.

2 Intellectual Property Protection and Plants

Sidney B. Williams, Jr.
Upjohn Company
Kalamazoo, Michigan

Kenneth A. Weber
Townsend and Townsend Law Firm
San Francisco, California

BACKGROUND

There is considerable discussion in both the USA and in international circles regarding the impact of providing utility (industrial) patent protection for plants. Prior to 1985, only limited protection was available for plant cultivars in the USA and elsewhere. With the issuance of utility patents in the USA, substantial new protection is afforded to plant breeders. Significant pressure now exists for the international community to permit patent protection for plants.

This chapter will discuss the legal basis for providing protection for plant cultivars in the USA and the relative value of the different forms of protection. The discussion will consist of an analysis of the three statutory schemes available for protecting plant varieties and related inventions. These laws include the Plant Patent Act (PPA), the Plant Variety Protection Act (PVPA), and the utility patent statute.

The chapter concludes with a discussion of the public policy aspects relating to the 1985 decision by the U.S. Patent and Trademark Office (Patent Office) to permit utility patents with claims to plants. Utility patents offer substantially broader protection for plants than was previously available under the PPA and PVPA. This important decision has, therefore, opened a vast new opportunity for the protection of agricultural research and development. Although it is the purpose of a strong intellectual property system to promote industrial development, significant debate exists concerning the direct impact of utility patents on the agricultural industry and their indirect impact on society.

Copyright © 1989 Crop Science Society of America, American Society of Agronomy, Soil Science Society of America, 677 S. Segoe Rd., Madison, WI 53711, USA. *Intellectual Property Rights Associated with Plants*, ASA Special Publication no. 52.

THE LEGAL FRAMEWORK

Three U.S. Statutes Provide Proprietary Protection for Plants

Plant Patents (35 U.S.C. 161 et seq.)

The original Plant Patent Law, the Townsend-Purnell Act, was enacted in 1930. Its purpose was to afford plant breeders the same type of protection for their inventions (plants) as that enjoyed by inventors of inventions having industrial application. The early 1900s saw several bills introduced into Congress but not passed into law. The proposed law permitted the patenting of products of nature and evoked considerable controversy. At the urging of such notable individuals such as plant breeders Paul Stark, Luther Burbank, and inventor Thomas Edison, and with the support of orchardman and Senator John Townsend, Congress provided patent protection for plants that are asexually reproduced.[1] (Refer to the Footnote section on p. 105.)

By plants, Congress meant the popular meaning of seed-bearing organisms (monocots or dicots). Bacteria and fungi are not patentable subject matter under the plant patent statute.[2] Asexual reproduction includes budding, grafting, and tissue culturing. The statute was amended in 1953 to provide protection for newly found plants (seedlings) once they are asexually propagated and placed into a cultivated state. Uncultivated and tuber propagated plants (such as Irish potatoes and Jerusalem artichokes) are excluded from protection.

There are now approximately 6000 issued plant patents. They are relatively short documents consisting of a written description of the claimed plant, a detailed colored illustration[3] (typically a photograph) and a single claim. The single claim uses a formalistic style such as: "We claim, a new and distinct variety of *genus, species* herein described and illustrated and identified by the characteristics enumerated above together with the parts thereof."

Prosecution of Plant Patents. The complete application consists of several parts; namely, a petition with appropriate fees, a specification with a single claim and an oath or declaration. Applications are filed in duplicate. A copy may be forwarded to the USDA-ARS, if the Patent Office requests a report from the Department regarding the novelty and distinctiveness of the claimed variety.[4]

Prosecution of plant patents is often relatively simple and straight forward. The following discussion will point out troublesome issues that may arise. There are two parts to the invention of a patentable plant.

The first part involves the recognition of a distinct phenotype as new.[5] The second part involves an actual reduction to practice by asexual reproduction. In the 1930s, the state of the art of horticulture was such that asexual means of reproduction was thought to be the only means to reliably reproduce a plant cultivar. Thus, our plant patent statute limits protection to those plants that can be successfully reproduced by cutting, grafting, or budding. Plants reproduced solely by sexual means are excluded from protection.

There is no constructive reduction to practice for plant patents. An actual reduction to practice by asexual reproduction is required and must be averred to in the inventor's declarations.[6] Because of these two parts to the invention of a patentable plant, the question of what constitutes inventorship becomes of interest.[7] Typically, the inventorship for a utility patent constitutes conception inclusive of a reduction to practice. Should the reduction to practice require inventive contribution by another, the possibility of adding that party as a co-inventor arises. In the typical plant patent fact pattern, asexual reproduction does not constitute an inventive contribution, yet it is arguably necessary to the conception. Whether to include the individual physically doing the asexual reproduction becomes a central question. To not investigate this issue is to place the enforceability of the subsequent patent at risk. To investigate requires an analysis of the facts surrounding the recognition of the distinct characteristics of the claimed plant, the discovery of novelty and the means for asexual reproduction. To be a co-inventor, it is probably necessary to have contributed to at least two of these three aspects of invention.

Another curious aspect concerning prosecution of plant patents involves the interaction between prior art and enablement. It is clear that prior art of 35 U.S.C. Sec. 102(a) and (b) are applicable to plant patents. However, to be effective art, the references must be enabling in that they must place the invention into the hands of the public. The typical written specification of a plant patent will not satisfy Sec. 112, the enablement requirements.[8] Thus, plant patents cannot anticipate plant patent applications unless the prior art patent provides a source for the claimed plant. This same logic applies for descriptions in monographs and other academic literature.[9]

A final aspect of prosecution involves the application of the nonobviousness requirement of 35 U.S.C. § 103. On a technical level, patented plants must embody at least one distinctive quality that is nonobvious to skilled breeders. This means that novel plant cultivars are not patentable if they merely express minor characteristic variations of color, taste, yield, or the like where those characteristics do not significnatly change the properties of the plant from its most similar phenotype. On the practical level, there is no readily articulable standard for the application of the nonobviousness requirement.[10] In practice it is a subjective requirement.[11] To avoid obviousness rejections it is best to include in the specification objective comparison data between the claimed variety and the closest phenotype known to the inventor.

Infringement. A plant patent provides narrow protection to its owner. The only protection a plant patentee has is the ability to exclude others from asexually reproducing the patented plant.[12] A patented plant cultivar reproduced by sexual means will not infringe a plant patent. Moreover, the derivation of a novel plant from a patented plant using its seed is not an act of infringement.

Unlike for utility patents, judicial opinions concerning the scope of protection for a plant patent are few and contradictory. The typical infringement situation arises when the patentee discovers a competitor selling a cultivar that is phenotypically similar if not identical to the patented cultivar. Evidence of a smoking gun, that is evidence of derivation from the patented plant may not always be present. If such evidence is available, liability is clear. In the absence of direct evidence or derivation, the case law is split. The majority of the judicial opinions require the patentee to present evidence of derivation. However, a minority hold that phenotypic identify is sufficient to establish infringement.[13]

When to File a Plant Patent. Although plant patents require asexual reproduction, all plants eligible for patent protection also reproduce by seed. Given the proper inbreeding, or appropriate self-pollination it is possible for several patentable plant cultivars to be reliably reproduced by sexual means as well as by asexual means. Where reliable seed are present, certification under the PVPA should also be considered.

Plant patents should be considered as a viable option where the breeder has not produced a stable seed source, has not identified a characteristic that is definable by Mendelian-type segregation and claimable per se, or where the applicant has limited interest in derivations or limited funds.

Plant patents are relatively cheap proprietary protection and although the scope of plant patent protection is narrow, they can be cost effective. This is especially true when a patentee expects to negotiate license agreements where specific patented cultuvars are to be sold and where the licensee has no interest in developing new cultivars to engineer around the patent.

Plant Variety Protection Certificates

The PVPA of 1970 provides patent-like protection in the USA for plants that are sexually reproduced (reproduced by seed).[14] It represents the culmination of an effort by many of those in the seed industry to provide incentives for private industry to conduct research and breed novel cultivars from seeds.

Plant cultivar protection is modelled along the lines of U.S. patent laws. It is administered, however, by the Department of Agriculture rather than the Patent and Trademark Office. Its purpose is to promote progress in agriculture by providing plant breeders with a limited exclusive right on new sexually reproduced plant cultivars in exchange for the plant breeders' disclosure of the cultivar and how to reproduct it.[15]

Prosecution. Standards of protectability are set forth in Sec. 2402(a) of the Statute.[16] All seed-bearing plants are eligible for protection. Bacteria, fungi, and first-generation hybrids are expressly excluded from coverage. The standards for examination are novelty and sexually reproduction.

To meet the standard of novelty, a cultivar must (i) not be barred by any of the activities set forth in Sec. 2402(a)(1), (a)(2), and (a)(3);[17] and (ii) be distinct, uniform, and stable.[18]

The statute defines the following as activity that will defeat novelty: (i) where the cultivar was a public cultivar before the applicant's date of determination; (ii) where the cultivar was a public cultivar more than 1 yr before the application; (iii) where the cultivar was available to workers and described in a printed publication in this country before the applicant's date of determination; (iv) where the cultivar was available to workers and described in a printed publication in this country more than 1 yr before the application for plant cultivar protection; (v) where the applicant or his proxy filed for protection in a foreign country for the same cultivar more than1 yr before filing in the USA;[19] or (vi) where another breeder has an earlier date of determination and has either obtained a Plant Variety Protection Certificate, has been continually engaged in development for commercialization, or has published a description of the cultivar within 24 wk of the date of determination.[20] These novelty provisions are analogous to the novelty provisions in the general patent statute, 35 U.S.C. § 102.

The other components of novelty, distinctiveness, uniformity, and stability are defined in Sec. 2401(a). For distinctiveness, the agency looks to any single characteristic that will impart novelty to a phenotype.[21]

Sexual reproduction (by seed) is another requirement of protectability. Sexual reproduction of the new cultivar must be averred to by the applicant.[22]

An application for a Plant Variety Protection Certificate consists of a completed application form (OMB No. 0581-0055) and accompanying exhibits. These forms and instructions can be obtained from the Plant Variety Protection Office.

The application must provide the name and address of the applicant (the actual breeders' names are not needed), a taxonomic description of the cultivar, the name of the cultivar or a temporary designation, a statement as to the novelty of the cultivar, a declaration averring compliance with the statute and that viable seed will be replenished if necessary, and the signature of the applicant or its representative.

The application form provides for the submission of five exhibits:

1. Origin and breeding history of the variety.
2. Novelty statement.
3. Objective description.
4. Addition description of the variety.
5. Basis of ownership.

The examiner may waive the necessity for submitting certain exhibits if it is determined that the information and descriptioncontained in other exhibits justify it.[23] Finally, the application must include at least 2500 viable seed along with $1800 for associated fees.

Three aspects of the formal application bear additional discussion. The application will specify the cultivar name. This is the permanent cultivar name and is analogous to the generic name of a chemical compound. Cultivaral denominations are an important aspect of the seed business, and it was one of the more hotly debated issues during the USAs deliberation as to whether or not to become a member of the International Union of the Protection of New Varieties of Plants (UPOV).

It is probable that the questions relating to cultivaral names will continue to generate much discussion. The purpose of the cultivar name is to identify the cultivar and not its source of origin. Therefore, it is suggested that cultivaral names should not be used as a substitute for a trademark. It may be prudent to consider marketing the cultivar under a second and registerable name (e.g., Blueboy™ cv. AA-100). A popular cultivar marketed under a trademark as well as its cultivaral designation will hopefully offer a competitive edge to the original owner beyond the 18-yr period of protection.[24]

The date of determination is the equivalent of an actual reduction to practice under the patent statutes. This is the date that a plant breeder first determines that his cultivar has certain characteristics.[25] These characteristics need not be new. The important factor is that the breeder should determine that the characteristics exist.

Plant Variety Protection Certificates differ from patents in that they need not be filed in the name of the breeder (inventor). They can be filed in the name of the owner of the cultivar.[26] If the applicant is a corporation, the application must contain the state and date of incorporation.

Applications for a Plant Variety Protection Certificate are filed with the Plant Variety Protection Office in Beltsville, MD. This office is a unit of the Livestock, Poultry, Grain and Seed Division, Agricultural Marketing Service of the USDA. Incomplete applications are held for 24 wk and the applicant is given appropriate notice. If no remedial action is taken within 24 wk, the application is considered abandoned.

To facilitate prosecution, it is advised that the application provide details as to novelty. Vague allegations will not support a certificate. If a particular characteristic is deemed critical to novelty, objective criteria should be offered to establish the distinctiveness of the phenotype.[27] The Plant Variety Protection Office may require actual plant specimens.[28]

Infringement. The PVPA offers significantly broader protection than the Plant Patent Act. A certificate gives its owner the right, inter alia, to prohibit others from selling or offering the protected cultivar for sale, importing or exporting the cultivar, sexually or asexually multiplying the cultivar.[29] In addition, there is a provision for infringement based upon the act of inducing others to directly infringe.[30]

The statute provides some exceptions to infringement. There is no infringement when sexual reproduction occurs in the context of developing a new inbred line and when asexual reproduction occurs "in pursuance of a valid United States plant patent".[31] The meaning of the latter provision is unclear.[32]

In addition, the PVPA expressly provides for four other exemptions.[33] These include: a grandfather exemption for persons having developed the identical cultivar more than 1 yr before a third party filed for certification; an exemption for farmers to sell to other farmers so long as the sale of seed does not become a primary business;[34] a research exemption for breeders developing new cultivars; and, an exemption for carriers and advertising businesses.

Remedies are statutory and provide for injunctive relief and money damages.[35] As with the patent statute, there is no recovery for acts of infringement committed more than 6 yr prior to filing for relief or for acts known to the owner for more than 1 yr before filing for relief.[36] Provisions for attorney fees are also available.[37] Actual notice is a requirement for obtaining damages.

Unlike the patent statute, the PVPA provides for recovery for acts of infringement committed prior to the issuance of a certificate. But similar to the patent statute, notice is required to obtain damages. Seed packaging can be suitably labelled to give warning of unauthorized propagation before issuance and after issuance of a certificate, the statute suggests the phrase, "U.S. Protected Variety".[38]

When to File for Plant Cultivar Protection. The PVPA can provide cost-effective protection for breeders. But it should be clearly understood that the protected cultivars can be modified in relatively small ways to leave infringement an open question. If given a choice between filing for a plant patent or a plant cultivar certificate, the greater protection afforded the owners of certificates makes certification the best choice.

Utility Patents for Plants (35 U.S.C. 100 et seq.)

In 1985, the Patent Office began issuing utility patents for all plants.[39] Prior to this time, the Patent Office issued utility patents only for first-generation plant hybrids because these plants were excluded from proprietary coverage by both of the specialty plant statutes. The utility patent is the most powerful proprietary protection available for plants and plant-related inventions. In direct correlation with the value and scope of protection, the requirements for obtaining utility patents are more stringent than those for obtaining grants under the special plant statutes.

Recalling that the special plant statutes provide protection only for the cultivar described in the patent or certificate, the plant utility patent is not so restricted. Certainly, cultivars can be claimed in a plant utility patent. However, in addition to cultivars, genera, and species, plant parts embodying some patentable character may be claimed if adequate support for patentability is provided.

Prosecution. The application and examination for a utility plant patent is identical with the application and examination of any utility patent.[40] By statute, any invention claimed in an enforceable utility patent must have (i) utility, (ii) novelty, and (iii) non-obviousness.[41] In addition, the application must place the invention in the hands of the public by describing its making and using.[42] Finally, the invention must be described in terms of its best mode contemplated by the inventors at the time of filing.[43]

To obtain the full value of the utility patent, it is helpful to have claims to a characteristic or select group of characteristics rather than simply to a phenotype. Where the underlying genetics have not been precisely characterized, it is helpful to describe the characteristic as a genetic factor.[44] Providing that the specification teaches how to obtain and transfer the genetic factor to other cultivars, the factor should be patentable without a detailed description of its genetic basis.

In general, utility and novelty are not problematic requirements to meet for plant utility patents. Any cultivated plant will most likely have utility and the novelty standards are similar to those established for plant patents. The two most notable requirements of the utility patent statute that distinguish it from the two specialty plant statutes are (i) that the application must contain a written description that enables one skilled in the art to reproduce and practice the invention[45] and (ii) that the invention be nonobvious (involve an inventive step) to one skilled in the art.[46]

Without the grace of 35 U.S.C. § 162 providing that "no plant patent will be declared invalid for noncompliance with Sec. 112 of this title if the descriptionis as complete as is reasonably possible," plant utility patents will generally require a deposit. Deposits are intended to ensure availability to the public. Applicants are not required to deposit until after allowable subject matter has been determined by an examiner.[47]

The Patent Office is currently promulgating new rules governing the deposit of biological material.[48] The rules are not yet finalized. Recently, a draft was made available for public comment. The proposed terms of the deposit are similar to the terms of the Budapest Treaty on the International Recognition of the Deposit of Microorganisms for the Purposes of Patent Procedure. The applicant will agree to maintain the deposit for at least 30 yr or 5 yr after the last request for deposit. In addition, the draft rules provide mechanisms for replenishing the deposit in the event of contamination or viability loss.

Deposit of plant material raises several practical problems. First, the selection of the form of deposit must be considered. Seed deposit can be the simplest, but with some plants, viability is a serious concern. The deposit may need to be replenished every 3 to 4 yr with seeds that lose 50% viability each year. When available, germplasm or cell culture are also viable alternatives.

The selection of a suitable depository is also of concern. The proposed Patent Office rules require specific conditions for a depository to be acceptable. These include continuous existence, independence, and competency.[49] The depository needs to be amendable to furnishing samples to the public in an "expeditious and proper manner." The expense of deposit can easily exceed the filing fees for a patent. The Federal seed depository in Fort Collins, CO has been an economical alternative for some clients. However, in view of the new proposed regulations, it is important to ensure that the depository will provide seed to the public if requested. Conversations with the personnel at the Fort Collins facility indicated reluctance on their part to be a primary source of seed for the public.

Finally, it is important to remember that deposit of seed under the Budapest Treaty may result in the availability of that deposit even if the underlying application is rejected, abandoned or withdrawn. Discussions are presently underway that hopefully will change this result,[50] but for now the publication of an application containing a deposit under the Budapest Treaty can result in unwanted release to the public.

The requirement of nonobviousness is a condition of patent grant. The purpose of the standard of nonobviousness is to ensure that patents will only be granted for those solutions to problems that involve skill over and beyond that possessed by the mythical person of ordinary skill in the art. Its presence or absence is determined by (i) what is known in the prior art, (ii) the difference between what is sought to be patented and the prior art, (iii) the level of skill of the average person in the art and (iv) secondary considerations or indicia of unobviousness.

The standard for nonobviousness appears to have been relaxed for plant patents.[51] When a court applied the above four elements for a plant patent, it found that the elements were inapplicable.[52] Applying the nonobviousness standard to plant breeding will require a determination of what are the closest or most similar cultivars known in the art to the cultivars for which protection is sought, and what is the difference or differences between the closest prior art cultivar and the cultivar for which protection is being sought. The *Yoder* court stated that the level of skill of the plant breeder with the species in question would not be meaningful criteria for a plant invention and it is unclear whether a court analyzing the obviousness of a plant utility patent would stress this third criteria or merely offer lip service to it.

After these three determinations are made, the fact finder is required to consider any other evidence of unobviousness. this other evidence, often referred to as indicia of unobviousness, includes evidence of (i) a long-felt need that has gone unsatisfied; (ii) the failure of others to solve the problem (i.e., resistance to a specific plant pathogen or drought resistance), (iii) copying the product, and (iv) commercial success. Commercial success involves more than showing impressive sales figures. It requires a showing that commercial success is attributable to the characteristic of the claimed cultivar. Each of the determinations is a factual one that will have to be decided with the help of expert testimony.

Recent cases involving the question of unobviousness have held that an Atlantic Ocean oyster is obvious in view of a Pacific Ocean oyster;[53] that the use of polyclonal antibodies in an assay does not render obvious the use of monoclonal antibodies in an assay;[54] and that a herbicide's properties need not produce a superior result in every environment.[55]

In the context of a plant, the typical obviousness rejection arises when the patent examiner is faced with a novel characteristic such as added color or increased sugar content and finds that such characteristic would have been one obvious to a skilled breeder to attempt to breed for the selected character trait using the known methods. The rebuttal generally consists of arguing that even if the phenotype was perhaps suggested, the underpinning genetics were unknown and, therefore, success could not have been predicted. For example, a character trait may be a complex of alleles that were not readily marshalled into a single individual. In addition, one can argue that the means to achieve the claimed individual or its phenotype are nonobvious and, therefore, the process can impart nonobviousness into the resulting product (plant). Finally, objective proof of failed attempts by others to achieve the claimed phenotype might sway the examiner that your plant is patentable.[56]

Scope of Protection. There are two types of infringement situations recognized in the USA, literal infringement and infringement under the doctrine of equivalents. Literal infringement occurs whenever a party has operated within the metes and bounds of your claim. By "operated" it is meant that there has been making, using, or selling of a claimed invention.

If the accused product of process does not fall within the literal term of the claims, there nevertheless can be infringement under the doctrine of equivalents. The doctrine of equivalents is an equitable solution to the problem arising where an activity or product embodies trivial differences from a claimed invention. The doctrine provides for findings of infringement where an accused produce or process accomplishes substantially the same function or result in substantially the same manner as a patented invention.

The infringement of a plant utility patent can occur as a result of the making, using, or selling of the plant or any derivation of that plant. With a dominating plant utility patent, a patentee can force others to take a license under an earlier patent even if a later developed cultivar is a patentable improvement over an earlier cultivar.

Unlike the special plant statutes, the utility patent has no express provisions for excluding methods of reproduction or for research use exceptions. Thus, the utility patent affords more protection to patentees than the other statutory forms of protection.

There is an exclusion to infringement wherein a patentee exhausts his patent rights following a first sale. In the traditional fact pattern, this exclusion permits third parties to sell a patented invention to a third party without infringing the patent. The exhaustion of rights after first sale has unique concerns where plants are being sold. The plant invention can be reproduced and if unrestricted the doctrine of exhaustion could render patent protection meaningless. Until the questions of exhaustion and implied licenses have been addressed by our courts, it is imprudent to release reproducible plant material (or any reproducing biologic) without contractual restrictions. Such limitations should include nontransferability provisions and restrictions upon reproduction.

In addition to the problem of exhaustion of rights, is the possibility of an implied license arising from the sale of a plant. An implied license typically arises in situations where a patented device is sold and the seller also holds a patent on a method of using the device. If the device has only a single intended use, a purchaser of that patented item has been granted an implied license to use that item for its intended use. A plant could be construed as being purchased for reproduction as well as for sale as a nonreproducible product (sale of fruit or leaf material). Specifically stated, the question is whether a grower who purchases a patented seed, would be granted an implied license to replant his field with the extra seed he harvests and does not sell as a food product.

When to File a Plant Utility Patent. Because the utility patents provide greater protection for a plant, one should always consider the possibility of filing a utility patent. The first practical consideration is whether the plant has a property that merits the broad protection afforded a utility patent. The more the invention is directed to a particular cultivar for its combination of characteristics with a reduced emphasis placed upon a specific characteristic that is transferrable to other cultivars, the less likely one will want to file a plant utility patent.

The cost of prosecuting and maintaining a plant utility patent is considerably greater than the cost arising from a plant patent or a plant cultivar protection certificate. A breeder having discovered a brighter rose color inclusive of other valuable characteristics might not care if others pursue this brighter color in distinct cultivars. Indeed the greater the focus is upon the cultivar as a whole, the harder it will be to establish the nonobviousness of the plant in the traditional context of a utility patent. Moreover, if the advantage begins sought through a utility patent is the prohibition of derivation, one needs to ask if the applicant would actually enforce its rights to limit such activity or whether the applicant could actually detect derivation should it occur.

On the other hand, where a breeder has discovered a specific trait that is transferable to other individuals and where that trait is a valuable characteristic, a plant utility patent may be appropriate and necessary. The choice of protection is a decision reflective of ones' focus. It is important to define the actual invention. Do you seek protection for a plant or a characteristic of the plant? The next question is whether the characteristic has been described in such a manner so as to meet the requirements for a utility patent.

If evidence of nonobviousness is marginal, it would be helpful to draft the plant utility patent specification in a manner that would also provide support for a plant patent. In the event that the statutory requirements for a utility patent are not met, a plant patent might prove to be a satisfactory second alternative.

Multiple Coverage under the Three U.S. Statutory Schemes

The question of obtaining two or more forms of proprietary protection arises from the existence of the three statutory systems described above. The three statutes provide different rights. These rights need not be viewed as in conflict, but as alternatives that complement each other. The existence of dual protection is not without precedent. Dual protection under design patents and copyrights or trademarks[57, 58] is available.

The principle objection to multiple forms of protection is the resulting extension of protection. The period of protection for a U.S. patent begins upon issuance and not at the filing date. Notwithstanding the relative advantages of each system, by permitting a patent to issue on a cultivar certified under the PVPA, one could extend protection for the cultivar well beyond the period deemed reasonable by Congress to promote the technical arts.

An express prohibition for double patenting is provided at 35 U.S.C. § 101.[59] This statute precludes a patentee from obtaining two patents for the same invention. Where nonobvious variations exist between two inventions in two separate applications owned by the same party, statutory double patenting is inapplicable; however, a judicially created double patenting doctrine requires the applicant to file a disclaimer of any patent period of the second patent that would otherwise extend beyond the patent life of the first patent.

The law of double patenting is applicable to plants. However, the strategy of filing a broad dominating patent on a plant having a specific characteristic or genetic factor, should not preclude a second patent issuing for a specific cultivar that expresses the characteristic claimed in the utility patent. This approach should prove of particular interest to breeders seeking plant patents for specific cultivars that might not be nonobvious in the traditional sense for utility patents.

The co-existence of a PVPA certificate and a plant patent is not expressly prohibited by statute. At least one author believes that the legislative history of the PVPA might preclude patent protection for plants that might otherwise be protected under the plant patent statutes.[60] It would, however, appear that the question is rather academic. Should a novel plant qualify for patent protection under either statutory scheme, the greater scope of protection under the PVPA should dictate the choice of proprietary protection.

There is consensus that simultaneous protection is available for plant cultivars that are both certifiable under the PVPA and patentable under U.S. utility patent statutes.[61]

HARMONIZATION OF U.S. LAW WITH INTERNATIONAL LAW

The USA is signatory of several international treaties governing intellectual property laws. The seminal treaty was the Convention of Paris for the Protection of Industrial property, 20 Mar. 1883. In the last 100 yr, the signator nations have periodically met and reaffirmed their commitment to a strong legal system for the protection of industrual property.

In 1961, seven of the convention members signed the International Convention for the Protection of New Varieties of Plants. This convention provided a uniform framework for proprietary protection for plants. Today, there are 17 members, including the USA. The USA adopted the UPOV convention by Presidential Executive Agreement.[62]

Membership in the Union requires harmonization of Member Nations patent or plant breeder's rights laws. At least four provisions of the convention made it necessary to consider amending some provisions of the U.S. law and/or Rules of Practice relating thereto. They are:

1. Requirement that a cultivar be given a denomination.[63]
2. Requirement that a sample of propagating material be supplied to the examining office on request.[64]
3. Loss of rights resulting from sale or marketing a cultivar.[65]
4. Experimental use.[66]

Concerning cultivaral denominations, the Plant Patent Act and utility patent statutes do not require that the application contain a name for the cultivar. However, many applications do contain cultivar names. In recognition of this conflict, proposed rules were published by the Patent Office in 1989 for public comment.[67] The PVPA rules are in compliance with UPOV.[68]

As part of its examination process, the Patent Office may require that an applicant submit a specimen of the variety. However, an applicant is not required to submit a sample upon filing the application, nor is a patent holder required to replenish propagating material. The PVPA requires a deposit of seed and is therefore in compliance with the UPOV.[69]

The novelty provisions of the Convention[70] are different from the novelty provisions of 35 U.S.C. § 102. However, Article 37 of the Convention allows the USA to apply the novelty provision of §102. The PVPA is in compliance with the UPOV provisions.[71]

Regarding experimental use, the UPOV convention does not require third parties to obtain permission of the cultivar owner before conducting research aimed toward developing new cultivars. There is no provision in U.S. patent statutes exempting experimental users from patent infringement liability. A judicially created experimental use exception has been recognized. The doctrine, however, is not universally followed and a recent Federal Circuit decision implies that there may be no experimental use exception for a profit-oriented entity.[72] In its most narrow interpretation, U.S. law may only find an exception where a patented invention is being tested for operability or where an invention is used in an abstract search for knowledge.[73] The research exemption of the PVPA brings that statute in compliance with the UPOV convention.

Finally, the members of UPOV have the option of either providing patent protection for plants or maintaining an independent system generally described as plant breeders' rights laws.[74] The USA has never acceded to this provision.

RECENT INTERNATIONAL DEVELOPMENTS

Recently the Commission of the European Communities has published a Proposal for a Council Directive on the Legal Protection of Biotechnical Inventions.[75] The proposal does not recommend a change to Article 53(b) of the European Patent Convention, which precludes patent protection for plant and animal varieties. Nevertheless, the Commission found no prohibition in the European Patent Convention for claims embracing protection for plants and animals where they are claimed generically, e.g., a tomato (*Lycopersicon esculentum* L.) plant comprising a gene encoding protein X.[76] In recommending that the nations of the European community formally approve generic patent protection for plants and animals, the Commission has adopted the positions already taken by the U.S. and Japanese patent offices.

Another interesting issue involves the Commission's recommendation that compulsory licenses (dependency licenses) be made available where plant breeders' rights conflict with patent rights. Compulsory licenses are already available in a number of European countries. The USA has historically opposed compulsory licenses except in exceptional situation.[77]

The Commission also recommended needed revisions to the deposit rules as they relate to access where applications have been abandoned or withdrawn. These recommendations have been previously discussed.[78]

CONCLUSION

The preceding overview of the statutory schemes available in the USA for protection of novel plant cultivars and related inventions illustrates the emphasis that the body politic placed on the importance of agriculture in our society. The fundamental purpose of these statutes is to encourage the development of an innovative and growing agricultural economy. The political authority arises from a Constitutional mandate to Congress to pass legislation that will promote the technical arts.

Utility patents afford vast new protection for plant breeders. Plant breeders who identify useful genes or create new genes can now obtain utility patents that will guarantee 17 yr of exclusive protection for that gene in any plant or derivation thereof. To use these exciting new genes, breeders will be asked to take a license. Typically, royalty payments are requested if marketable products are developed through the use of patented plant product. In some instances, licenses may be exclusive and not available to all competitors.

Utility patents have been especially effective for industries that require long periods for product development at high costs and produce products that will be marketed for long periods. The food, chemical, and pharmaceutical industries are typical examples of industries that extensively use and rely upon the patent system.

With the availability of utility patents, the seed and nursery industries will face new challenges. There is opportunity for growth and development. There is the possibility that inefficient companies will not survive. Utility patents will permit and perhaps create change. The agricultural industry in America has been handed an exciting new opportunity.

To promote the technical arts is the objective of any statutory scheme protecting industrial property. The question of whether any particular scheme achieves its intended purpose is a complicated question that goes beyond the scope of this discussion. The debate over whether a patent system actually benefits society or limits opportunity and growth has engaged academicians for more than a century. The failure to empirically answer the problem lies in the lack of an effective model to study the relative benefits of an intellectual property system. We rely instead on the legislative process and the power of the individual to change the system.

The USA is a leader in agricultural research and sets the standards for international statutory schemes. The law is a reflection of our world. Both are in a constant state of change. The debate we are currently engaged in is both exciting a significant. For it is through such discussions that society will determine the scope of plant proprietary protection. As we balance economic and social concerns arising from exclusive intellectual property rights to agricultural research and development, we should keep in mind that the objective of the law is to afford the greatest benefit to society.

FOOTNOTES

1. Doyle, J. 1985. *Rodale's organic gardening.* Designs on nature, Part 1. p. 24–32, Nov.
2. *In re Arzberger,* 112 F. 2nd 834, 46 U.S.P.Q. 32 [(Court of Custom and Patent Appeals), 1940].
3. 35 U.S.C. § 165.
4. 37 C.F.R. § 163(b).
5. 35 U.S.C. § 161.
6. Patent and Trademark Office Rule 62 requires the inventors to aver that the claimed plant has been asexually reproduced.
7. The intentional deletion or naming of a noninventor to a patent application may render any subsequent patent unenforceable.
8. 35 U.S.C. § 162 relaxes the Sec. 112 requirements for an enabling disclosure.
9. *In re Le Grice,* 301 F.2d 929; 133 U.S.P.Q. 365 (CCPA, 1962).
10. *Yoder Bros v. California-Florida Plant Corp.* 537 F.2d 1347 at 1378–1379; 193 U.S.P.Q. 164 (5th Cir. 1976).
11. Novel plants claimed in plant patents appear to be viewed as nonobvious unless prior art suggesting the specific combination of traits is found. This is a difficult test to meet and obviousness rejections are uncommon during prosecution of plant patents.
12. 35 U.S.C. § 163.
13. Bernstein, D.B. 1986. Is a plant patent a form of copyright? Idea (Spring) (1):31–38.
14. 7 U.S.C. § 2321 et seq.
15. 7 U.S.C. § 2581 and the Plant Variety Protection Act, preamble: An Act to encourage the development of novel varieties of sexually reproduced plants and to make them available to the public, providing protection available to those who breed, develop, or discover them and thereby promoting progress in agriculture in the public interest.
16. 7 U.S.C. § 2402.
17. 7 U.S.C. § 2402(a)(1), (a)(2) and (a)(3).
18. 7 U.S.C. § 2401(a).
19. See also 7 C.F.R. 180.7 (7) wherein the rules provide for an extension of time for foreign applicants. This rule was promulgated to assure compliance with the UPOV convention.
20. 7 U.S.C. § 2402(a).
21. See *In re Walker,* 40 Agric. Dec. 1017 (1981), ruling that the novelty requirement is met where a single characteristic is distinct, uniform, and stable.
22. 7 C.R.F. § 180.7(a).
23. 7 C.R.F. § 180.5(b)(2).
24. This dual approach is useful regardless of the particular proprietary protection being selected.
25. 7 U.S.C. 2401(d): The term *date of determination* means the date when there has been at least tentative determination that the cultivar has been asexually reproduced with recognized (sic) characteristics, whether or not the novelty of those characteristics has been determined.
26. 7 U.S.C. § 2421(a): An application for a Certificate of Plant Variety Protection may be filed by the owner of the cultivar sought to be protected. The application shall be made in writing to the Secretary, shall be signed by or on behalf of the applicant, and shall be accompanied by the prescribed fee.
27. *In re Jack B. Fleming,* 38 Agric. Dec. 917 (1979), requiring verifiable scientific data in support of an application for plant certification.
28. 7 C.F.R. § 180.8.
29. See 7 U.S.C. § 2541 Infringement of plant variety protection.
30. 7 U.S.C. § 2541 (8).

31. 7 U.S.C. § 2541 (7) provides: [It] shall be an act of infringement . . . to perform without authority . . . (7) perform any of the foregoing acts even in instances in which the novel variety is multiplied other than sexually, except in pursuance of a valid United States plant patent; . . .
32. See Chisum, *Patents* § 1.05[5] note 45.
33. 7 U.S.C. §§ 2542-2545.
34. See *Delta & Pine Land Co. v. Peoples Gin Co.,* 694 F.2d 1012 (5th Cir. 1983) holding that the farmers' exemption of § 2543 does not permit a farm cooperative to actively solicit buyers on behalf of a farmer.
35. 7 U.S.C. §§ 2563–2564.
36. 7 U.S.C. § 2566. This 1-yr provision engenders interesting issues for parties learning of infringement prior to the issuing of their certificate.
37. 7 U.S.C. §§ 2565.
38. 7 U.S.C. § 2567.
39. *Ex parte Hibberd et al.,* 227 U.S.P.Q. 443 (PTO BD. Pat. App & Int. 1985). *Hibberd* held that plant patents and plant cultivar protection certificates were not exclusive forms of protection for plants and therefore plants are protectible under the utility (general) patent law. The Hibberd application involved maize having increased tryptophan levels and it contained claims to seeds, whole plants, and tissue cultures.
40. In brief, a patent application will consist of a specification including at least one claim. Accompanying the application are a transmittal form with the appropriate filing fee and formal papers which include a formal declaration by the inventors, assignments for recordation in the Patent Office, a Small Entity Statement if appropriate and a Power of Attorney.
41. 35 U.S.C. §§ 101–103.
42. 35 U.S.C. § 112.
43. 35 U.S.C. § 112.
44. Claim 20 of U.S. Patent No. 4 513 532: A genetic factor derived from corn, which genetic factor is capable of conferring an extra leaf phenotype which is capable of transmission to progeny substantially as a single dominant gene.
45. 35 U.S.C. § 112.
46. 35 U.S.C. § 103.
47. *In re Lundak,* 773 U.S.P.Q. 1216 (Fed. Cir. 1985). It should be noted that deposit rules of many foreign nations require that the deposit be perfected in accordance with the Budapest Treaty prior to the filing of the priority application in order to obtain an effective filing date.
48. Official Gazette, 1095 OG 10-22 25 Oct. 1988.
49. The specific requirements are detailed in the proposed rules at Sec. 1.202. 1095 OG 20 25 Oct. 1988.
50. Proposal for a council directive on the legal protection of biotechnical inventions, commission of the European communities. Chapter 4: Proposed article 15 deposit, access and redeposit. Brussels, 17 Oct. 1988.
51. A more rigorous analysis under § 103 is currently being considered by the Patent Office for utility plant patents than is presently applied to plant patents. (personal communication, see also note 11 and accompanying text.)
52. *Yoder Bros.* Supra at Note 537 F.2d at 1378–1379.
53. *In re Allen,* 2 U.S.P.Q. 2d 1425 (Bd. Pat. App. & Int. 1987) (1983).
54. *Hybritech Inc. v. monoclonal Antibodies, Inc.,* 231 U.S.P.Q. 81 (Fed. Cir 1986).
55. *In re Chupp,* U.S.P.Q. (Fed. Cir. 1987).
56. See Bent et al. 1987. *Intellectual property rights in biotechnology worldwide,* M. Stockton Press, New York at 192–195 for an interesting discussion of prosecution strategy for plant type utility patents.
57. *In re Yardley,* 181 U.S.P.Q. 331 (CCPA, 1974).
58. *In re Mogen David Wine Corp.,* 140 U.S.P.Q. 575 (CCPA, 1964).
59. Sec. 101 provides, whoever invents or discovers any new and useful process machine, manufacture or composition of matter . . . may obtain a patent for . . . [emphasis added].
60. Juergensmeyer and Wadler, *Agricultural Law,* Vol. 1, § 18.3.4 Little Brown and Co. (1982).
61. Neagley, Jeffery and Diepenbrock, "Section 101 plant patents—panacea or pitfall", 1 APLA selected papers A-1 (1983) and Adler, R.G. Can patents coexist with breeders' rights? Developments in U.S. and International Biotechnology Law. IIC 17(2):195–227 (1982).

62. The Congress has not ratified this treaty and, therefore, the USA is not bound to the terms of the treaty where our national laws are in conflict. *In re Hibberd,* 227 U.S.P.Q. 443 [Patent and Trademark Office Board of Appeals and Interferences (PTOBAI), 1985].
63. Article 6(e) of the UPOV Convention: The cultivar shall be given a denomination in accordance with the provisions of Article 13.
64. Article 10 of the UPOV Convention: All three U.S. statutory schemes provide for deposit and access by an appropriate agency to test the applicants' plants.
65. Article 6(b)(ii) of the UPOV Convention.
66. Article 5(3) of the UPOV Convention: Authorization by the breeder shall not be required either for the utilization of the cultivar as initial source of variation for the puspose of creating other cultivars or for the marketing of such cultivars. Such authorization shall be required, however, when the repeated use of the cultivar is necessary for the commercial production of another cultivar.
67. 52 Fed. Reg. 42016.
68. 7 U.S.C. § 2422. An application for a certificate recognizing plant variety rights shall contain: (1) The name of the variety except that a temporary designation will suffice until the certificate is to be issued.
69. 7 C.F.R. 180.6(d). The applicant shall submit with the application at least 2,500 seeds of the viable basic seed required to reproduce the variety.
70. Art. 6 of the UPOV convention.
71. See 7 C.F.R. 180.7 and Christie, A. The novelty requirement in plant breeders' rights law, IIC Vol. 19(5):646–657 (1988).
72. *Roche Products v. Bolar Pharmaceuticals Co.,* 221 U.S.P.Q. 937 (Fed. Cir. 1984) and *Eli Lilly v. A.H. Robins,* 228 U.S.P.Q. 757 (Dist. Ct. E.D. Va. 1985). Both these cases involve acts directed to commercializing products identical to a patented product. A case involving acts directed to the improvement using a patented product might result in an opinion establishing an experimental exception relating to research and development.
73. See Hantman, R.D. 1985. Experimental use as an exception to patent infringement. JPOS 67(12): 617–644.
74. UPOV Art. 2(1): Forms of protection: Each member . . . may recognize the right of breeder . . . by the grant either of a special title of protection or of a patent. Nevertheless, a member State of the Union whose national law admits of protection under both these forms may provide only one of them for one and the same botanical genus or species.
75. See Note 48 supra.
76. See Note 48 supra at p. 33–35 and Article 3. The Commission appears to have clearly affirmed the 1983 opinion of the EPO Technical Board of Appeal in *In re Ciba-Geigy,* 1984 Official Journal EPO 112. It is, however, unclear whether the Commission would extend patent protection for plants obtained by traditional selective breeding methods.
77. The eminent domain provision of 28 U.S.C. 1498 and the public interest in wide usage provision of the PVPA 7 U.S.C. 2404.
78. See Note 48 supra and the accompanying text.

3 Patenting Genes that Encode Agriculturally Important Traits

Stephen A. Bent
Foley & Lardner, Schwartz, Jeffery, Schwaab, Mack, Blumenthal & Evans
Alexandria, Virginia

Where intellectual property rights are concerned, the international viewpoint has historically been that biological creations in general, and the embodiments of agricultural advances in particular, fall outside the ambit of patent laws deemed properly applicable to "industrial" inventions only (Bent et al., 1987). It is only fairly recently that a revolution in the way we think about biological systems has forced lawyers, judges, and government officials throughout the world to reassess their preconceptions of where innovation in applied biology, including the agricultural sciences, fits into the framework of laws intended to foster technological progress. Even in western Europe, where widespread experimentation with the patenting of plant cultivars over a quarter-century ago was stymied by the establishment of the current, exclusive system of plant-cultivaral protection, there are now persistent calls for statutory reform by those who believe that proprietary rights in plant and animal inventions should be on par with those relating to inanimate creations (Baeumer, 1987).

Advocates for change in this area often refer with approval, and not a small amount of envy, to the situation in the USA, where several options are available for protecting plant-related inventions, and even animals are patentable (Bent, 1989). Ironically, opponents to these vehicles for proprietary rights—and a well-entrenched opposition there is, make no mistake—likewise invoke events in the USA to argue for their position. The assertion is commonly made, for example, that U.S. law has come to the point where large corporations are able to patent the "genes of agriculture" and, thereby, achieve the "monopoly" power to control food production, gobble up traditional breeding concerns, obliterate the family farm, or do all of the above (Doyle, 1986).

Rhetoric of this sort raises questions regarding societal values—the intrinsic worth of family farms, for instance—that transcend the scope of this presentation. On the other hand, we can reasonably inquire whether the basic scenario underlying this concern, namely, the extensive patenting of agriculturally important genes per se, is a realistic one.

Copyright © 1989 Crop Science Society of America, American Society of Agronomy, Soil Science Society of America, 677 S. Segoe Rd., Madison, WI 53711, USA. *Intellectual Property Rights Associated with Plants*, ASA Special Publication no. 52.

To gain some insight on this issue, let us compare the cases of two issued U.S. Patents, no. 4 535 060 (Comai, 1985) and no. 4 581 847 (Hibberd et al., 1985), directed to inventions that are based, respectively, on the alleged discovery of a new genetic determinant for a valuable agronomic characteristic. The subject matter claimed in each patent relates to a biosynthetic pathway, common to aromatic amino acids like tryptophan, phenylalanine, and tyrosine, which encompasses many different enzyme-catalyzed reactions (Fig. 3–1). More specifically, the Hibberd patent concerns mutant maize (*Zea mays* L.) material that overproduces tryptophan, while the Comai patent pertains to a mutant gene encoding a herbicide-resistant variant of one of the enzymes of the pathway depicted in Fig. 3–1.

TRYPTOPHAN OVERPRODUCTION IN MAIZE

The first patent we shall examine issued on an application that was the subject of a widely noted decision, *Ex Parte Hibberd*, in which the U.S. Patent and Trademark Office (examining authority), by way of its Board of Appeals, stated that both utility patent protection and USDA-type cultivaral protection for plant-related inventions are available under U.S. law.[1] The patent included claims of the following sort:

1. A maize plant capable of producing seed having an endogenous free tryptophan content of at least about one-tenth milligram per gram dry seed weight, wherein the seed is capable of germinating into a plant capable of producing seed having an endogenous free tryptophan content of at least about one-tenth milligram per gram dry seed weight.
2. A tryptophan overproducing maize plant capable of expressing a gene coding for an anthranilate synthase which retains from about 60 to about 80 percent of its activity in the presence of 5-methyltryptophan at a concentration of about 10^{-5} *M*, wherein the plant is capable of transmitting the gene encoding the anthranilate synthase to its progeny.
3. A maize plant derived from seeds deposited with In Vitro International, Inc. and assigned Accession No. 10010.

Sets of claims also issued to maize seed, a hybrid maize plant, and a maize tissue culture, each set incorporating terms corresponding to those employed in Claims 1, 2, and 3 above, respectively.

The applicants had originally submitted claims that were not limited to maize, but instead had broadly recited "mutant monocot seed," including seed of a "a cereal crop" such as "maize, rice, wheat, barley, sorghum, oats, rye [and] millet." In the application as filed, the claimed invention was said to provide "for the production of a plant cell enzyme that is relatively resis-

[1] 227 U.S.P.Q. 443 (Bd. Pat. App. & Int., 1985).

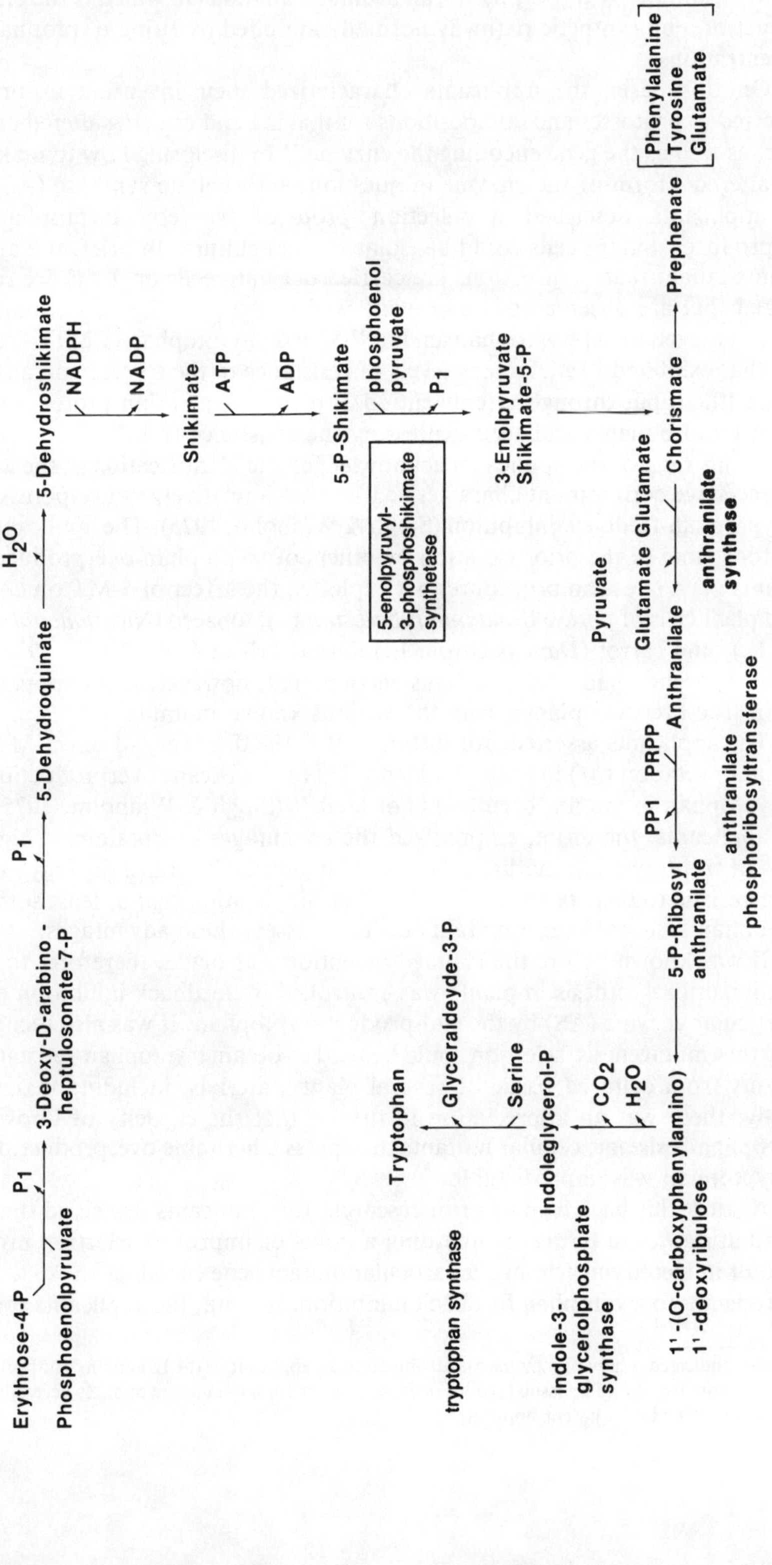

Fig. 3–1. Flowchart for biosynthesis of aromatic amino acids (Adapted from Umbarger, Ann. Rev. Biochem. 47:533, 574–576 [1978]).

tant to inhibition by tryptophan," an aromatic amino acid which is the end product of a biosynthetic pathway normally inhibited by rising tryptophan concentrations.[2]

On this basis, the applicants characterized their invention as one "directed to processes and compositions for making and utilizing altered enzyme, as well as the gene encoding the enzyme." In disclosing how to make an "altered" form of the enzyme in question, anthranilate synthase (AS), the applicants described a selection protocol whereby tryptophan-overproducer mutant cells could be isolated from culture. In brief, the applicants taught that "[s]elections are carried out until cells or tissue are recovered that are observed to be growing well in the presence of normally inhibitory levels of [a] tryptophan analog," 5-methyltryptophan (5-MT). Cell lines that exhibited "satisfactory levels of resistance to the tryptophan analog are [then] put through a [conventional] plant regeneration protocol to obtain mature plants and seed expressing the resistance trait."

In this regard, the applicants acknowledged the identification in the art of a recessive corn mutant characterized by an AS relatively nonresponsive to tryptophan-feedback inhibition (Singh & Widholm, 1975). The applicants also took note of the prior isolation by others of tryptophan-overproducer mutants via a selection procedure that exploited the affect of 5-MT on cultured plant cells of potato (*Solanum tuberosum* L.), tobacco (*Nicotiana tabacum* L.), and carrot (*Daucus carota* L.) lines (Carlson & Widholm, 1978). There apparently had been problems encountered, however, in attempts to obtain true-breeding plants from the various known mutants.

The applicants asserted, for instance, that the discovery of one 5-MT-resistant mutation (*bf*) in maize had "not led to significant overproduction of tryptophan in mutant corn plants or seeds" (Singh & Widholm, 1975). The applicants, therefore, emphasized the advantages of obtaining "high levels of tryptophan in mature plants. . .and genetically transmit[ting] this characteristic to their progeny," with a clear implication that at least some tryptophan-resistant maize mutants do not possess these advantages.

It was known before the claimed invention was made, therefore, that tryptophan biosynthesis in plants was controlled by feedback inhibition of a particular enzyme (AS) by the end-product tryptophan. It was also recognized that mutagenesis/selection could be used to obtain tryptophan-resistant mutants from cultured tissue of several plant materials, including maize. Finally, there was an appreciation in the art that the capacity of known tryptophan-resistant, cellular mutants to express a heritable overproduction of tryptophan was unpredictable.

Against this backdrop of prior research, the applicants described their contribution *not* in terms of providing a novel or improved selection protocol, or in identifying/cloning a particular mutant gene encoding an AS that was resistant to tryptophan feedback inhibition. Instead, the applicants spe-

---

[2]Unless indicated otherwise, statements attributed to an applicant or the Examining Authority are taken from the text of an issued patent or from a paper contained in the publicly accessible prosecution record of a patent application.

cifically disclosed the isolation and manipulation of a particular corn mutant, derived from cultured tissue of a particular hybrid corn line, that possessed the above-mentioned advantages of regenerability and heritability. Based on progeny studies of plants regenerated from this mutant, the applicants also determined that the expressed tryptophan-resistance trait was "meiotically transmissible as a dominant character and behave[d] as would be expected for a [Mendelian] trait encoded by a single nuclear gene."

But the applicants also asserted that other types of tryptophan overproducers, including those reflecting "single or multiple site [gene] mutations [or] the presence of multiple gene copies of the enzyme," came within their claimed invention. By the same token, plant material genetically engineered to express exogenous DNA encoding an enzyme "that catalyzes the same reactions as" the altered anthranilate synthase described in the application were deemed to be covered by the submitted claims. As a corollary, only certain of the claims, such as Claim 3, were limited by a recitation of a mutant expressly described in the applicants' working example and represented by biological material placed by them on public deposit.

When the examining authority first took the Hibberd application up for consideration, broad claims were presented covering tryptophan-overproducing "monocotyledonous seed," including seed containing "an anthranilate synthase gene that does not naturally occur in the seed," such as a gene "from a prokaryotic organism." The examining authority rejected all claims, in particular for an alleged lack of "an enabling disclosure." Contending that those of ordinary skill, given what the application taught, would be unable to create the broad range of plant materials embraced by these claims, the examining authority alleged that "[u]ndue experimentation would be involved in (i) the selection of appropriate donor tissues, (ii) establishing suitable media and culture conditions, (iii) the selection of effective mutagens, (iv) cell line selection, (v) plant regeneration, and (vi) establishment of a breeding strategy." Even the claims limited to maize were considered overly broad in light of the fact "that only two cell lines of tryptophan overproducers were isolated, [both]. . .from a single variety of maize." Claims embracing "genetic manipulation, which is not found in the [applicants' working] example," were also specifically rejected.

In effect, the examining authority took the position that the applicants were entitled to claims only as broad as their deposited material, generated in accordance with their working examples (see Claim 3). The applicants responded by canceling all of the originally considered claims and replacing them with claims substantially identical to those eventually issued, that is, claims limited to maize.

To overcome the lack-of-enablement rejections, they argued that the "individual procedures" of selecting appropriate donor tissue, establishing suitable culture conditions and the like were "novel in their combination and application as disclosed in the specification, [and could] be applied with certainty," or at least "with a degree of experimentation that is commensurate with the nature of the invention and the state of the art." To substantiate this position, the applicants submitted the declaration of one of their number, Dr. Kenneth Hibberd, detailing the isolation of cellular mutants, and the regeneration of mutant plants, from "tryptophan overproducing tissue culture cell lines in two additional maize genotypes." According to the applicants, this declaration demonstrated that their patent disclosure enabled those of ordinary skill to obtain "maize tissue cultures and plants that overproduce tryptophan, and seeds that contain increased levels of tryptophan," from "a wide variety of maize genotypes."

In the face of the applicants' arguments and supporting declaration evidence, the examining authority withdrew the nonenablement rejection, leading to the eventual issuance of claims that the applicants asserted in the record were still broad enough to read on *any* maize "mutant, recombinant or genetically engineered derivative thereof having the recited property relating to tryptophan biosynthesis." The applicants thus avoided the claim-limiting impact of their germplasm deposit by convincing the examining authority that the disclosed selection process, itself adapted to maize from the earlier work of others, could be employed with reasonable success, using well-defined selection criteria, in generating mutants within Claims 1 and 2 from a large number of different corn cultivars.

In addition to their showing that tryptophan-overproducer mutants would be derived from three different sources of corn tissue, the applicants' position was supported by evidence that *one* mutant described in the working example of their application was the result of a change at a single genetic site or "locus." This observation is consistent with the fact that a known tryptophan-resistant corn mutant had also been localized to a single Mendelian determinant, albeit a recessive one.

It is at least arguable that one of ordinary skill would expect such a "single-shot" mutation to be repeatable, under the same selection conditions, within a limited number of cell generations. In this regard, it must be kept in mind that genes can differ in mutability over several orders of magnitude. The relative stability (or instability) of the gene(s) involved in producing a mutant with certain characteristics would be an expected factor in determining the ease with which other mutants, similarly characterized, could be generated de novo. Conversely, the capability to generate de novo a tryptophan overproducer obtained by "multiple site mutations," absent any information about the relative location of the multiple loci involved, the frequency of mutation at the different loci, or the number of gene copies available for mutation, would be uncertain, to say the least.

From this perspective, a claim directed to maize material that expresses a dominant tryptophan-overproduction trait encoded by a single gene would seem to conform more closely to what the applicants' actually enabled in the Hibberd case. By the same token, the applicants' failure to disclose means for isolating and cloning *a* gene encoding an altered, tryptophan-resistant anthranilate synthase (not to mention an undisclosed enzyme catalyzing "the same reaction" as anthranilate synthase) undercuts the contention that the issued claims are enabled to the extent of covering plants genetically engineered to overproduce tryptophan.

GLYPHOSATE RESISTANCE

A common scenario in biotechnological innovation entails the transferal (by whatever means) of a genetic determinant that is identified, in functional terms, by reference to some aspect of the physiology of a particular organism or type of organism. In this scenario, the patent applicant typically must overcome the argument that a person of "ordinary skill" in the relevant art, once alerted to the existence of a given organic process or phenomenon in a particular type of living matter, would have considered it obvious to isolate and exploit the genetic information that drives the process or phenomenon.

In the Hibberd case, the applicants may have considered this argument to be a telling one and concluded, therefore, that prior disclosures concerning tryptophan-overproduction in maize and other crops precluded claims to a genetic determinant coding for that trait. In other words, they seem to have decided that a claim to "a gene encoding the overproduction of tryptophan" (or language to that effect) would not be distinguishable over prior art indicating the existence, at least, of such a gene.

In the Comai case, by contrast, a different conclusion was reached, as reflected by the presence, both in the original application and in the issued patent, of claims to "a DNA sequence of less than about [5000 nucleotide base-pairs] having a structural gene coding for a glyphosate resistant 5-enolpyruvyl-3-phosphoshikimate synthetase," which we shall refer to hereafter simply as "synthetase." The applicant in the Comai case thus claimed an exclusive right to any DNA sequence of a certain length that contained a gene characterized by its information content alone.

"Glyphosate" is the popular chemical name for one of the world's most widely used herbicides, n-phosphonomethyl glycine, which is sold under the trademark Roundup (Monsanto Co., St. Louis). Before the claimed invention was made, it had been known that glyphosate inhibited aromatic amino acid synthesis in plants and bacteria by interacting with the recited synthetase to the exclusion of the normal substrate of that enzyme. Because of competition between glyphosate and the normal substrate, the synthetase-catalyzed production of phosphoshikimic acid—an important precursor of aromatic amino acids like tryptophan—is disrupted, killing the target organism.

The applicant described in his patent disclosure a selection process that could be applied to mutated cells in culture to isolate an altered form of the synthetase structural gene encoding an enzyme that did not bind glyphosate preferentially to the normal substrate of the synthetase. Although the applicant's working examples involve the isolation of glyphosate-resistant mutants of the bacterium *Salmonella typhimurium*, the applicant stated that the DNA sequence encoding the glyphosate-resistance trait "may come from either prokaryotes or eukaryotes," including "bacteria such as. . .Escherichia, fungi, . . .yeast, plants [and] algae." The location of the mutated gene had been pinpointed at one of a series of genetic sites ("loci") corresponding, respectively, to the known positions (designated "*aro*A," "*aro*B," "*aro*C" and so forth) of genes involved in the synthesis of aromatic amino acids.

According to the applicant, the first step in isolating and cloning "the mutagenized [gene at the] *aro*A locus" involved mutagenesis of a prokaryotic or eukaryotic "host" organism that was glyphosate sensitive, that could be grown in vitro, and that had "the capability of expressing the. . .synthetase structural gene to provide the desired enzyme." Recognizing that glyphosate resistance among the variant population produced by mutagenesis could "result [from] a number of different types of changes in the [host] cell," the applicant also described the method for establishing "that the mutation had occurred in the *aro*A gene." With this method, the applicant stated that he had been able to obtain glyphosate resistant *S. typhimurium* mutants, using a known chemical mutagen, with a frequency of 1×10^{-4} per cultured cell, with "about 1% of all mutations conferring glyphosate resistance mapped close to, or in, [the] *aro*A locus."

Two separate grounds were cited for rejecting original claims in the application. In the first instance, the examining authority argued that the claimed "cell containing glyphosate resistant. . .synthetase" was not new, but rather was "anticipated by spontaneous mutations that would occur naturally." In this regard, the requirement of the claim that the mutation imparting glyphosate resistance must originate "from in vitro mutation" was discounted as "patentably irrelevant" because the mutation end product, whether arising spontaneously or otherwise, would be the same.

The claimed invention was also rejected on the ground that inventive step was lacking, in that the claimed invention would have been obvious in light of what had been known previously. More specifically, the examining authority argued that the broadest original claims were unpatentable over art showing "a structural gene coding for mutated enzyme [which is]. . .resistant to *histidine* inhibition" (emphasis added).

Thus, the examining authority took the position that prior knowledge in the art of *another* "mutated enzyme," itself modified so as to be no longer responsive to feedback inhibition by histidine, would have suggested *an analogy* "to applicant's synthetase that is resistant to *glyphosate* inhibition." In other words, rather than relying on a known phenotype to infer the existence of a genetic determinant, the examining authority analogized from the known behavior of one genetic system (*his*) to that of another system (*aro*).

In the ensuing dialogue, the applicant argued vigorously that the *possibility* that the recited mutation had occurred spontaneously at the *aro*A locus was "totally irrelevant [to patentability of the claims] if no one [had previously] isolated [the mutation], characterized it and reported its existence." An earlier, but unpublished, decision by the PTO Board of Appeals had, in fact, held that the *theoretical* occurrence of a spontaneous mutation was not the proper basis for a lack-of-novelty rejection.[3] Although that decision

[3] *Ex Parte Blair*, Appeal no. 525-532 (23 Sept. 1983), regarding U.S. application serial no. 205 022, now U.S. patent no. 4 444 888 (issued 24 Apr. 1984).

was not cited in the glyphosate-resistance case, the examining authority eventually abandoned the anticipation rejection.

To overcome the obviousness rejection, the applicant had to dissuade the examining authority from the position that manipulation by mutagenesis in the *his* system would have suggested another manipulation by mutagenesis in a different genetic system. It was argued, therefore, that modification of the known *his* gene "so that [the enzyme encoded by the gene] may no longer be. . .regulated cannot be generalized to the proposition that any gene can be mutated to obtain an expression product having a particular property." In support of this position, it was pointed out that the known gene mutation resulted in a modification of the encoded enzyme "at a site which is distant from the active site of enzymatic activity," at "a binding site that is involved with allosteric inhibition of the enzyme, whereby. . .binding [of histidine] to this allosteric site changes the conformation of the enzyme."

In contrast, the applicant urged, the recited "in vitro mutation in the aro locus" unexpectedly altered the catalytic site of the encoded synthetase "so that the ratio of binding between the substrate of the enzyme and the glyphosate [is] substantially modified. . .while still retaining a substantial degree of the natural enzymatic activity." The applicant thus invoked structural and related functional differences separating the respective enzymes of the *his* and *aro* systems to support his position that a mutation having a specific effect in one system would not have suggested a similar mutation in the other system.

But the examining authority maintained the obviousness rejection, arguing that the applicants had applied "an analogous method" to one previously "performed on a different gene," and that the known "mechanism of glyphosate inhibition" would have made it obvious to utilize "that inhibition. . .to isolate mutants resistant to [glyphosate]." Claims to a cell culture "consisting essentially of cells" containing the glyphosate resistance gene were eventually allowed, however, after the applicant specified in his broadest cell-culture claims that the glyphosate inhibition achieved with his invention was "less than about one-half of [that observed]. . .prior to mutation." This limitation was not incorporated, however, into the "DNA sequence" claims, which issued with relatively little specific discussion about them appearing in the official record.

DISCUSSION

In many ways, the flow of argument and counter argument in the Comai case is typical of disputes over the patentability of biotechnological inventions. On one hand, the examining authority (or accused infringer, in an altogether different context) contends that the starting material for the claimed invention would have been expected to embody sufficient variability to accommodate the production, via a known variability potentiating process, of a desired variant for which identifying selection criteria are available or can be readily designed. The typical rebuttal assertions are (i) that it would not have been obvious that the starting material could be modified to produce the variant in question and/or (ii) that design of selection criteria permitting the isolation of the desired variant would not have been a routine endeavor for those of ordinary skill.

Assertion (2) was not available to the applicants in the Hibberd case, nor could they have argued their primacy in discovering the genetic correlate(s) of tryptophan overproduction, in maize or any other organism. Yet they obtained claims to products (maize plant and seed) that are defined in terms of a chemical constituency (tryptophan content) or a physiological parameter (enzymatic resistance to 5-methyltryptophan), without reference to a particular genetic determinant. Thus, Claim 1 ostensibly covers *any* maize plant yielding seed meeting the prescribed free-tryptophan requirements, *regardless of the underlying genetic basis for the high-tryptophan trait.*

It remains to be seen whether the actual contribution made by the patentees in the Hibberd case would justify a court's finding infringement of Claim 1 by a maize plant expressing high free-tryptophan values via a mechanism wholly distinct from the one implicated in the Hibberd application. Nevertheless, it is significant to the present analysis that the patentees, at least as evidenced by statements made in the record of the Hibberd case, believe the protection afforded their invention to be undiminished by the fact that no "gene" per se is even recited, let alone claimed, in any claim of the Hibberd patent.

By contrast, the patentees in the Comai case took the position that they had been the first to isolate, in a reproducible manner, and to characterize a particular mutated gene—a gene "at the *aro*A locus" encoding a glyphosate-tolerant form of as—and, moreover, to discover that the altered gene could be used to obtain glyphosate tolerance in plants. (Objective evidence on this last point was submitted during prosecution of the Comai application.) Under these circumstances, where the genetic determinant involved was "engineered" (via a mutagenesis technique) or, more generally, had not been "known" to exist prior to applicants' discovery, the examining authority concluded that absolute patent protection of the determinant, i.e., protection without limitation as to context, was warranted.

An analysis of these two U.S. patent cases prompts several observations relevant to the question of whether widespread patenting of agriculturally important genes is a likely prospect. First, the very question may be misplaced in its emphasis, since protection can be secured, in principle, under patent claims that are unfettered by recitations of a genetical nature and, hence, are potentially *broader* than claims directed to a particular determinant. This fact also reminds us that biological inventions are typically amenable to description at several levels—molecular, cellular, organismic, and physiological—and that a judicious choice of descriptive level in a patent application can pay dividends in terms of claim scope.

Secondly, one is hard pressed to think of many traits of even theoretical interest to agriculturalists that are not already "known," in a patent-legal sense, to be associated with a particular genetic underpinning. The discovery of a determinant imparting glyphosate-resistance represents a notable exception in this regard, as might the isolation, say, from a native cultivar, of a gene for a trait unheralded in the art, such as resistance to a disease engendered by a new (or newly mutated) pathogen.

Accordingly, genes that are in some way "synthetic" or that code for a previously unrecognized trait, and hence fall into the "patentable-in-principle" category, seem destined to comprise a distinct minority among the genetic determinants significant to agriculture. Conversely, plant-related inventions distinguished over the art solely on the basis of quantitative traits such as yield and vigor, which generally reflect the interaction of many, non-clustered genes, are likely to remain beyond the reach of absolute protection, at least until we gain facility with transferring, as a heritable unit, artificial assemblages of interacting genes into a desired background.

On the other hand, no general technological or legal impediment stands in the way of patenting a distinct group of plants—a "variety" or "line"—sharing certain traits that are predictably observed from generation to generation and which differentiate the plants of the group from a "typical" form of the species, as well as from other possible cultivars within that species. Indeed, such claims have issued in several U.S. patents, each of apparently narrow scope by virtue of incorporating, if implicitly, an open-ended list of characterizing traits that can be met ("enabled," in patent jargon) only by recourse to some manner of biological deposit which represents the invention itself.[4]

Given a range of patenting formats that stretches from claims covering specific DNA molecules to cultivaral claims encompassing whole constellations of genes, what should be our attitude on the appropriate role patent protection is to play, relative to plant breeders' rights and trade secrecy, with regard to "agrogenetic" technology? The key to answering this question lies in an appreciation of a "self-limiting" aspect of modern patent systems which fairly mandates that patenting should remain but one option for protection of intellectual property rights in this area.

Why this is so is illustrated in Fig. 3–2, where a universe encompassing all innovations in a given technological field—say, virus resistance in plants—is depicted in three "dimensions." A fourth dimension, that of time, is accommodated by illustrating three different stages in the evolution of the technology. The respective "spaces" of these stages have been offset, one from the other, for the sake of clarity; the differences in volume have no particular meaning.

The ordinate of Fig. 3–2 represents a crucial ratio in our conceptual universe, that of patentable-to-unpatentable subject matter (PUP), while the abscissa defines a second dimension in terms of the enabling disclosure that is technically feasible. As one moves away from the origin in the xy-plane, the value of PUP increases as a function of the increasing number of inno-

[4]See, for example, the claim in U.S. patent no. 4 594 810 (issued 17 June 1986) to a yellow dent corn "inbred. . .line having the designation HBA1," represented by germplasm deposited, on behalf of DeKalb-Pfizer Genetics, under accession no. 40225 at the American Type Culture Collection in Rockville, MD.

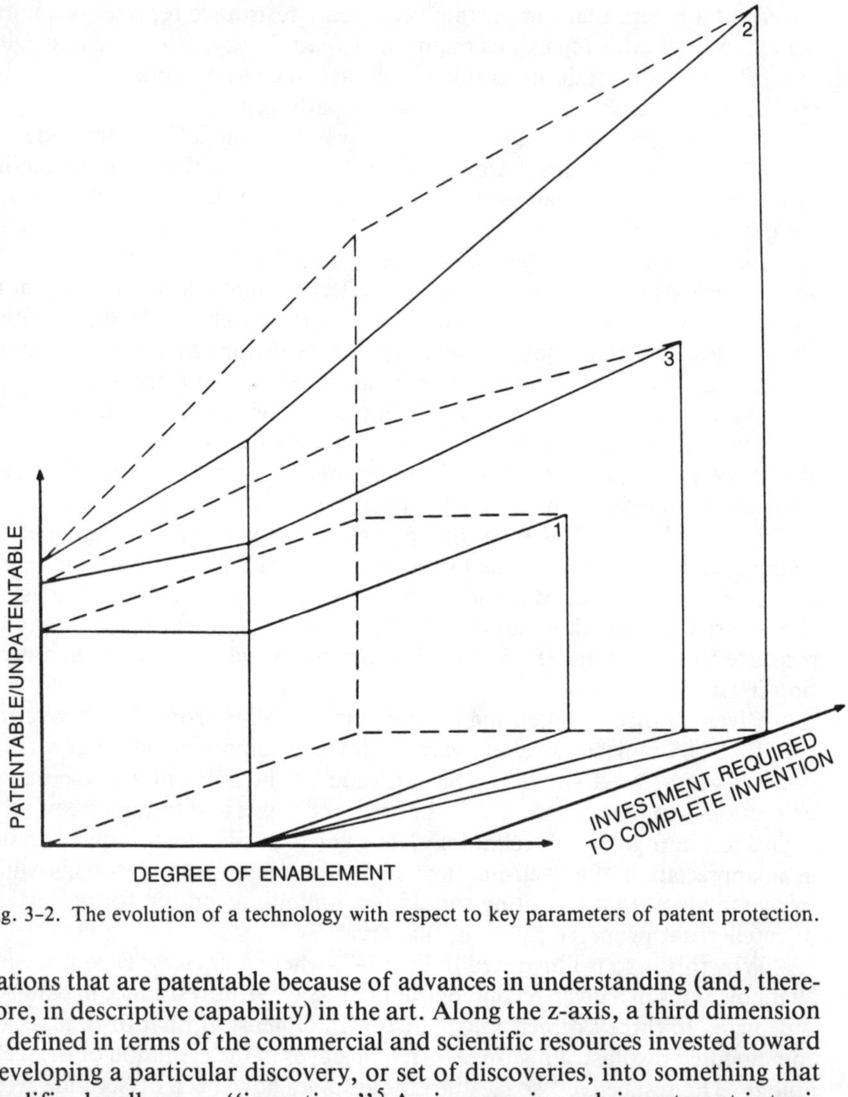

Fig. 3-2. The evolution of a technology with respect to key parameters of patent protection.

vations that are patentable because of advances in understanding (and, therefore, in descriptive capability) in the art. Along the z-axis, a third dimension is defined in terms of the commercial and scientific resources invested toward developing a particular discovery, or set of discoveries, into something that qualifies legally as an "invention."[5] An increase in such investment is typically associated with an increase in PUP—greater expenditures in research generate greater numbers of patentable inventions—as well as in knowledge that can be applied to the task of providing an enabling disclosure.

[5] Virtually all patent systems distinguish in some way between "discoveries" and "inventions."

> A "discovery" in a patent law sense is an abstract idea, an intangible conception, e.g., a natural force, a law of nature or natural phenomenon, a property of a substance, a mathematical relationship or the like. One cannot patent a mere discovery; something more is needed in order to turn a "discovery" into an "invention." The necessary element is the *application* of the discovery to a utility falling within the concept of. . . [the] "useful arts". . .

Bent et al. (1987) Chapter 4 at note 19 (citations omitted).

In situation 1 of the figure, we find one theoretical extreme of our model: a scenario where the value of PUP stays the same regardless of the resources expended or the depth of the disclosure furnished. This situation could hardly pertain in even the most mature of technologies.[6] By contrast, the dynamics of an immature technology are depicted in situation 2, where expenditures in research and development are more likely to yield pioneer inventions warranting protection that is broad in scope. By the same token, an innovation under the circumstances of situation 2 will typically occur in a technological environment that is relatively "uncrowded," in the sense that there are fewer prior developments that will have to be distinguished over to satisfy patent legal requirements for novelty and inventive step. Thus, vectors in either the xy- or xz-plane are sharply positive in situation 2, a state of affairs that is representative of much of agricultural biotechnology today.

But a crucial point to realize is that expansive situation 2 *must* eventually give way to situation 3, where the ever-increasing mass of general information, coming in part from the patent disclosures realized out of an earlier technological stage, is reflected in a decline in the rate at which PUP increases with increased expenditures and descriptive capability in the art, respectively. As more and more becomes known about the organization and properties of particular types of living matter (say, plants that display cross-protection against different viral pathogens), the harder it becomes to exploit that living matter in a manner that is not reasonably predictable a priori, no matter how much money is spent and detailed description provided. Moreover, the scope of protection that is available, even in principle, becomes more and more stringently delimited by the patent practitioner's difficulty in drafting claims that can be interpreted broadly without reading on an expanding prior art.

Put another way, Fig. 3–2 shows how the conventional standards for patentability act in concert with the principles governing patent-infringement determinations to secure a niche for a spectrum of living-matter compositions that will always lie beyond the reach of cost-effective patent protection. Some of these compositions will be reserved to public usage by operation of the product-of-nature doctrine, or by virtue of their having been developed out of publicly funded research. Still others will fall within that region of Fig. 3–2 reserved to valuable biological inventions for which a patenting prerequisite, such as nonobviousness or enabling disclosure, is considered out of reach.

Trade secrecy will be the vehicle of choice for commercializing some of these creations. For the rest of the valuable but unpatentable inventions, the availability of the protection afforded under the Plant Variety Protec-

[6] Nevertheless, it was the *status quo* effectively presumed, circa the late 1950s, by those who wrote into the laws of Western Europe the exclusions from patentability that continue to plague biotechnologists possessed of vastly expanded understanding of living matter and state-of-the-art descriptive techniques. It was assumed, in other words, that certain kinds of multicellular organisms and "essentially biological" processes, could (or should) not be patented for all times, regardless of future advances in the relevant arts.

tion Act will remain an important factor favoring their dissemination and improvement. Accordingly, there will *always* be a need and a demand for the specially tailored protection scheme of plant breeders' rights, existing *in parallel with* the traditional patent and trade-secrecy laws, to accommodate the full range of biotechnological innovations, including the development of agriculturally important genetic constructs.

REFERENCES

Baeumer, L. 1987. Protection of inventions in the field of biotechnology. p. 15–26. *In* Symposium on the protection of biotechnological inventions. Proc. Symp. WIPO and Cornell Univ., Ithaca, NY. 4–5 June. WIPO Publ. 657(E), Geneva.

Bent, S.A. 1989. Animal patents: Issues and prospects in the United States. *In* Animal patents: The legal, economic and social issues. Macmillan, London. (In press.)

Bent, S.A., R.L. Schwaab, D.G. Conlin, and D.D. Jeffery. 1987. Protecting property rights in biotechnology worldwide. Macmillan, London.

Carlson, J.E., and J.M. Widholm. 1978. Separation of two forms of anthranilate synthetase from 5-methyltryptophan-susceptible and -resistant cultured *Solanum tuberosum* cells. Physiol. Plant. 44:251–255.

Comai, L. 1985. Inhibition resistant 5-enolpyruvyl-3-phosphoshikimate synthetase, production and use. U.S. Patent 4 535 060. Date issued: 13 August.

Doyle, J. 1986. Altered harvests. Viking Penguin, New York.

Hibberd, K.A., P.C. Anderson, and M. Barker. 1985. Tryptophan overproducer mutants of cereal crops. U.S. Patent 4 581 847. Date issued: 15 April.

Singh, M., and J.M. Widholm. 1975. Study of a corn (*Zea mays* L.) mutant (*Blue fluorescent-1*) which accumulates anthranilic acid and its β-glucoside. Biochem. Genet. 13:357–367.

Troyer, A.F. 1986. Inbred corn line. U.S. Patent 4 594 810. Date issued: 17 July.

4 Protecting Plant Germplasm: Alternatives to Patent and Plant Variety Protection

Jeffrey L. Ihnen and Robert J. Jondle

Robbins & Laramie
Washington, DC

The development of plant germplasm for man's use has been ongoing for millenia, with the most rapid and notable advances in the last 70 to 100 yr. Significant advances have been made in the development of higher-yielding hybrids and cultivars, as well as improvements in insect and disease resistance. Other advantages have resulted from the utilization of chemical fertilizers and biocides, and the emerging area of biotechnology.

The use of biotechnical techniques in the development of plant germplasm is becoming more commonplace. For example, it is now possible to fingerprint plant cultivars using restriction fragment-length polymorphisms (RFLP) or two-dimensional (2-D) gel electrophoresis. Using these techniques, the relationships of genotypes can be analyzed and their degree of relatedness determined. The RFLP can also be used to track genes in a breeding program, and in certain instances, improve research efficiency. It may soon be possible for these techniques to be used to reverse engineer plant cultivars, i.e., determine the genetic message of a plant cultivar (a trade secret) by starting with, and analyzing, the plant cultivar.

The development of plant germplasm requires a significant investment, in both time and money, by either a breeder or a biotechnologist. Effective protection of plant germplasm is desirable, to either (i) ensure recovery of the investment and provide incentive for further research in the development of new germplasm, or (ii) ensure widespread use of the germplasm.

Several types of protection for plant germplasm are available in the USA, including (i) utility patents,[1] (refer to the Footnote section on p. 141) (ii) plant patents,[2] (iii) plant variety protection (PVP) certificates,[3] (iv) trade secrets, (v) contracts, and (vi) the law of conversion. Utility patents were first authorized in the USA by the Patent Act of 1790. The current Patent Act of 1952 is directed to "anything under the sun that is made by man."[4] Plants were first held to be patentable subject matter under 35 U.S.C. 101 in *Ex Parte Hibberd*.[5] The Plant Patent Act (PPA) was enacted in 1930 to provide patent protection to plants that are asexually reproduced. In 1970,

Copyright © 1989 Crop Science Society of America, American Society of Agronomy, Soil Science Society of America, 677 S. Segoe Rd., Madison, WI 53711, USA. *Intellectual Property Rights Associated with Plants*, ASA Special Publication no. 52.

the Plant Variety Protection Act (PVPA) was enacted to provide patent-like protection (under the auspices of the Department of Agriculture) to sexually reproducible plant cultivars.

Which type of protection will be most effective for any given germplasm will depend on several factors, including (i) the scope of protection desired, (ii) the type of germplasm to be protected, and (iii) the relative ease in producing the germplasm. This relationship can be illustrated as follows. A sexually reproduced hybrid cannot be protected by either a PVP certificate or a plant patent (PPA). A cultivar that is not reproduced asexually cannot be protected by a plant patent (PPA). A plant part cannot be protected by either a plant patent (PPA) or a PVP certificate. The patent statutes, in contrast to the PVPA, do not include a farmer's exemption that allows the reproduction and sale of protected seed. Once plant germplasm becomes publicly available, it may no longer be protectable by trade secret. It can readily be seen, therefore, that numerous factors must be considered in selecting the appropriate protection for plant germplasm.

The present paper addresses alternatives to patent or plant variety protection for the protection of plant germplasm, primarily on the basis of property rights in, i.e., ownership of, the germplasm. Alternatives also exist that may provide protection to an owner in the sense of protecting his market position, and as such, would have some commercial value. One of these alternatives is a trademark.

A trademark is used to identify the source of the trademark owner's goods, products, or services. If a high-value product (e.g., a specific cultivar or hybrid) is associated with a particular trademark, the owner of that product should be able to maintain his market position as a result of brand recognition and loyalty.

A second alternative for protecting market position is an action for a false designation of origin under Section 43(a) of the Lanham Act.[6] This alternative was utilized by Pioneer Hi-Bred International in a recent case,[7] in essence to recover damages from Holden's Foundation Seeds, Inc. for a reduced market share. In this action, Pioneer alleged that Holden's, in its applications for PVP certificates, did not identify the inbreds in suit as being derived from Pioneer's material. Pioneer also alleged that Holden's violated the Lanham Act by its promotional activities in connection with descriptions of the breeding, research, and development of the inbreds. It was further alleged that Holden's knew (or should have known) that its inbreds were derived from Pioneer's material, and hence, falsely designated the origin of the inbreds to create a likelihood of confusion, deception, or mistake on the part of the public. The court agreed with these allegations under Section 43(a), stating:

> The parties would agree and the evidence shows that Holden's LH38 and LH39 are not identical to Pioneer's H3H or H43SZ7. Holden did make some changes in its developments of LH38, LH39, and LH40 and, therefore, at least technically it was not a false representation for Holden to say in its marketing and elsewhere that it had developed LH38 and LH39. However, this Court is persuaded that Holden explicitly represented that LH38 and LH39 were developed

in its breeding program, and thereby implicitly represented to its customers and the trade that the parent material of LH38 and LH39 did not include any of Pioneer's private lines. Holden misled the public by not stating that LH38 and LH39 were derived from H3H and/or H43SZ7 and instead held out to the public then, and still does now, that LH38 and LH39 were developed solely by Holden by the use of L120, represented to be a Holden proprietary inbred.[8]

Since patent protection and plant variety protection are the subjects of other chapters in this series, the present chapter will address their alternatives, such as trade secrets, contracts, and the law of conversion. This chapter will examine strengths, weaknesses, advantages, and disadvantages of alternative protection for plant germplasm, relative to patent and plant variety protection.

It is important to remember that patent protection and plant variety protection are based on federal statutes, while all of the alternative protection for plant germplasm is based on state statutes or common law. It is not possible to address the issues on a state-by-state basis in this chapter, but a general analysis of the alternatives to patent protection and plant variety protection will be given. Although this chapter has been divided into sections for analysis purposes, there is much interplay between each type of protection as well as between patent protection and plant variety protection.

TRADE SECRETS

A *trade secret* is defined as any information that gives its owner an advantage over competitors who do not have knowledge of that information. The information is acquired in the course of the owner's business.A trade secret owner can prevent a third party from using or disclosing a trade secret if it was obtained illegally (i.e., misappropriated).

Trade secrets can provide both an effective and efficient means of protection for commercially sensitive and important business information. Numerous rationales have been given to explain the need for trade secret law, including (i) an incentive to innovate,[9] (ii) commercial morality,[10] (iii) sanctity of contract, (iv) enforcement of confidential relations, and (v) protection of property. Trade secret law can protect technology, although the degree of protection afforded by trade secrets is dependent on the technology itself. This section of this chapter will provide a general review of trade secret law and discuss its applicability in the protection of plant germplasm.

Definition

To protect proprietary information, a company must possess a trade secret in order to avail itself of trade secret law. What is a trade secret? Although there is no single definition, there are two widely accepted ones. The first is set forth in Section 757, Comment b, of the *Restatement of Torts* (1939) (hereinafter referred to as the *Restatement*). Comment b states, in part, that a trade secret

> . . . may consist of any formula, pattern, device, or compilation of information which is used in one's business, and which gives him an opportunity to obtain an advantage over competitors who do not know or use it. It may be a formula for a chemical compound, a process of manufacturing, treating or preserving materials, a pattern for a machine or other device, or a list of customers.

This definition prevails in the majority of state jurisdictions and has been adopted by the U.S. Supreme Court.[11] According to the *Restatement* definition, in order to qualify as a trade secret, information must (1) be used in one's business, (ii) provide a competitive advantage, and (iii) be secret.

A second widely accepted definition of a trade secret is set forth in Section (1) (4) of the Uniform Trade Secrets Act (UTSA), which defines a trade secret as:

> . . . information, including a formula, pattern, compilation, program, device, method, technique, or process that (i) derives independent economic value, actual or potential, from not being generally known to, and not being readily ascertainable by proper means by, other persons who can obtain economic value from its disclosure or use, and (ii) is the subject of efforts that are reasonable under the circumstances to maintain its secrecy.

This definition has been adopted in at least 17 states, and possibly as many as 22 states at present.[12] It is interesting to note, however, that the UTSA does not require that a trade secret be used in one's business, and thus is a bit broader than the *Restatement* definition. The UTSA definition can also protect as trade secrets subject matter that may not be protected under the *Restatement* definition.[13] However, the two definitions are generally consistent, in that they require (i) a competitive advantage be imparted by the information, and (ii) the information must be secret.

Subject Matter

It is clear in both of the above definitions that, unlike patents and PVP certificates, there is no specific subject matter criteria for a trade secret.[14] Almost any information can be a trade secret[15] as long as the requisites of secrecy, competitive advantage, and use (*Restatement*) are met. When the requisite criteria have been met, the following types of information have been found to be trade secrets:

> . . . ingredients in foods, drugs, and cosmetics; chemical compositions; computer programs and data bases; manufacturing, technological, and scientific processes; geological and geophysical maps; flowcharts and blueprints; product specifications; customer lists; employee, training, or other company manuals; and virtually any body of information.

Recently, a U.S. district court in Iowa held that a genetic message of an inbred corn (*Zea mays* L.) line can be proper subject matter for a trade secret.[16] Specifically, the court stated:

> In light of the totality of the circumstances, the Court holds that the genetic messages of H3H and H43SZ7 are "trade secrets." The genetic message of these lines of corn which Pioneer spent a great deal of money and effort developing is akin to a secret formula. This valuable formula did not exist outside of Pioneer's

> fields and the fields of its contractors and could only be duplicated with a great deal of effort and some luck. Furthermore, the evidence revealed that Pioneer took all reasonable precautions to protect the secrecy of the genetic message of these lines.[17]

Thus, plant germplasm is proper subject matter for trade secret protection as long as the requirements of use, competitive advantage, and secrecy are met.

Use and Competitive Advantage

In establishing the existence of a trade secret, the requirements of use and competitive advantage are minimal, while secrecy is by far the most important. Use and competitive advantage are often deemed to exist merely by virtue of a company's taking steps to meet the secrecy requirement.[18] In fact, it may be more appropriate to consider use and competitive advantage merely as criteria rather than actual requirements, since very little case law has been devoted to a discussion of them. In essence, these criteria require that the designated information (such as plant germplasm) be believed reasonably to provide the trade secret owner with some nonspeculative value and benefit in order to qualify as a trade secret.[19]

Secrecy

While meeting use and competitive advantage requirements is arbitrary at best, the secrecy requirement must be met without exception. In general, if subject matter is not actually secret, it will not be protectable as a trade secret. Comment b, Section 757 of the *Restatement* sets forth the most frequently cited standard for secrecy, and states in part that:

> The subject matter of a trade secret must be secret. Matters of public knowledge or of general knowledge in an industry cannot be appropriated by one as his secret. Matters which are completely disclosed by the goods which one markets cannot be his secret. Substantially, a trade secret is known only in the particular business in which it is used. It is not requisite that only the proprietor of the business know it. He may, without losing his protection, communicate it to employees involved in its use. He may likewise communicate it to others pledged to secrecy. Others may also know of it independently, as, for example, when they have discovered the process or formula by independent invention and are keeping it secret. Nevertheless, a substantial element of secrecy must exist so that, except by the use of improper means, there would be difficulty in acquiring the information.

A trade secret has only to be maintained in relative secrecy. The secrecy does not have to be absolute, i.e., known only to its owner. The secrecy must be considered to be reasonable, and in essence to be such that the trade secret is confided only in those who need to know in order to commercially exploit it. It may be necessary to disclose a trade secret to employees, contractors, suppliers, coventurers, and the like, in order to commercially utilize the trade secret.[20] However, any disclosure should be confidential, or made under conditions that imply confidentiality. Generally, the standards of secrecy can be examined by analyzing two fundamental elements: (i) the general knowledge or availability of the information, and (ii) the presence of affirmative steps to protect the information.

A first fundamental element of secrecy is that the information should not be generally known or available. If the subject matter of the trade secret has been voluntarily disclosed by the owner,[21] or is public knowledge or general knowledge in the trade or business,[22] it can no longer be protected as a trade secret. In essence, then, it is likely that secrecy does not exist for any information that has been published or distributed in any way, such as that contained in a book, magazine, trade journal, advertising brochure, or catalog, or in a patent or any other public format. Similarly, if the product itself discloses the trade secret, the product must remain confidential. It is this requirement, in fact, that may foster ineffectual protection of plant germplasm by trade secret. Public information, however, may be protected as a trade secret if the value or utility of the information is not obvious or is not otherwise known to competitors.[23] The information may be independently developed by several companies and still be considered a trade secret for each of them.[24]

A second fundamental element of secrecy is that the trade secret owner must take adequate affirmative steps to protect the trade secret information. In general, a company must take greater care to protect its trade secret than it would take to protect other aspects of its business. If a company believes that such a trade secret exists, then active steps must be taken to maintain its secrecy and confidentiality. If such steps are not taken, trade secret protection will be lost.[25] The degree of secrecy does not have to be absolute, but it must be reasonable.[26]

Several general principles can be enumerated that should be considered in determining the affirmative steps to take in order to maintain secrecy. First, trade secret information and related documents should be available to employees and other individuals on only a need-to-know basis.[27] Second, a balanced security approach is more important than stressing only one or two aspects.[28] Third, a company must realistically identify information that should be considered trade secrets, and then determine what protective steps are necessary. An excellent guide in determining such protective steps is the standard practice in each industry.[29]

Several steps that should be taken by a company in the protection of proprietary information include the following:

1. Establish a written trade secret policy.
2. Inform employees of trade secrets and remind them of their obligations of confidentiality.
3. Restrict access (by physical separation, if possible) to trade secrets.
4. Institute physical security measures (including storing trade secret documents in locked, limited-access files).
5. Protect trade secret documents by labeling them clearly with proprietary notice.
6. Maintain computer secrecy.

7. Restrict company tours.
8. Screen speeches and publications for unnecessary disclosure of trade secrets.
9. Inform others of the confidentiality of any disclosed trade secrets.[30]

Misappropriation

Trade secret law provides the trade secret owner with the right to prevent the unauthorized use and disclosure of a trade secret by other parties. As the U.S. Supreme Court said in *Kewanee*.[31]

> The protection accorded the trade secret holder is against the disclosure or unauthorized use of a trade secret by those to whom the secret has been confided under the express or implied restriction of non-disclosure or non-use. The law also protects the holder of a trade secret against disclosure or use when the knowledge is gained not by the owner's volition but by some "improper means." *Restatement of Torts* § 757(a), which may include a theft, wiretapping, or even aerial reconnaissance. A trade secret law, however, does not offer protection against discovery by fair and honest means such as by independent invention, accidental disclosure or by so-called reverse engineering, that is, by starting with a known produce and working backwards to divine the process which aided in its development or manufacture.

Section 757(a) of the *Restatement* states that disclosure or use of a trade secret constitutes misappropriation if (i) the trade secret was discovered by improper means; (ii) the disclosure or use of the trade secret breaches a duty of confidence to the party from whom the secret was obtained; (iii) the party was on notice that the secret was acquired by improper means or through breach of duty to another; or (iv) the party was on notice that the trade secret was mistakenly disclosed.

In Section (1) (2), the UTSA defines *misappropriation* to be:

(i) Acquisition of a trade secret of another by a person who knows or has reason to know that the trade secret was acquired by improper means; or

(ii) Disclosure or use of a trade secret of another without express or implied consent by a person who:
   (A) Used improper means to acquire knowledge of the trade secret;
   (B) At the time of disclosure or use, knew or had reason to know that his knowledge of the trade secret was (I) derived from or through a person who had utilized improper means to acquire it; (II) acquired under circumstances giving rise to a duty to maintain its secrecy or limit its use; or (III) derived from or through a person who owed a duty to the person seeking relief to maintain its secrecy or limit its use; or
   (C) Before a material change of his position, knew or had reason to know that it was a trade secret and that knowledge of it had been acquired by accident or mistake.

Improper means includes theft, bribery, misrepresentation, breach of duty (or inducement to breach) to maintain secrecy, trespass, or espionage.[32]

An otherwise lawful means that evidences an intent to appropriate a trade secret without permission, independent invention, or discovery may also be considered to be improper.[33] A confidential relationship can be established by contract, such as nondisclosure agreement or a post-employment restrictive convenant, or by an implied contract/special relationship that establishes that the disclosure was made in confidence and was not to be disclosed or used. A party can be put on notice that the information is a trade secret, and that the disclosure was the result of a breach of confidence by (i) actual notice (i.e., the party is expressly informed by the trade secret owner), or (ii) constructive notice (i.e., the facts and circumstances are such that the party should have known the information was a trade secret and was obtained through a breach of confidence).[34] In the absence of actual or constructive notice, a party is not liable to the trade secret owner.[35] However, the party may be liable to the trade secret owner after receipt of such notice, unless the party has materially changed positions and paid value for the secret in good faith before receiving notice.[36] The burden of proving misappropriation rests with the trade secret owner. If this burden is initially met, then the onus is shifted to the alleged misappropriator to prove that the trade secret was obtained by legal means.[37]

Although trade secret law protects an owner from the misappropriation of its trade secret, it does not protect against the discovery of the trade secret by proper means. Proper means of discovering a trade secret include independent discovery, accidental or actual disclosure, and reverse engineering.[38] If a product discloses the trade secret, or the trade secret can be determined from an analysis of the product, then the marketing of the product will result in a loss of trade secret protection. Trade secrets can also be lost by disclosure in patents, published patent applications, or other published materials, including advertisements or product brochures. With trade secrets that may need to be disclosed to governmental agencies, it is critical to indicate their confidentiality in order to avoid an accidental disclosure of the trade secret as a result of Freedom of Information Act (FOIA) requests. If the trade secret can be readily obtained from public documents, trade secret law may not provide protection, depending on the jurisdiction in which the misappropriation occurred, against a party who misappropriates it.[39]

Applicability to Plant Germplasm

The above discussion points out the criteria necessary for identifying and maintaining trade secrets, and the ability of trade secret law as an effective mechanism for protection of proprietary information. Although it is clear that plant germplasm is proper subject matter for trade secrets, the usefulness of trade secret law in the protection of plant germplasm is much less clear, especially in view of recent technological advances in the characterization of plant germplasm. The application of biotechnology to the development of plant germplasm will result eventually in the ability to reverse engineer essential characteristics of any given plant cultivar. It may remain difficult, however, to completely reverse engineer a given cultivar X to produce an identical variety, but it will be possible to reverse engineer a cultivar Y that will have the essential characteristics (such as yield, disease resistance, combinability, seed content, and the like) of cultivar X.

The fact that plant germplasm is living matter makes analysis of the applicability of trade secret law to its protection all the more difficult. It is clear that the sale of a product that discloses a trade secret results in the loss of that trade secret, because the product can then be reverse engineered by analyzing its composition or structure. However, the sale of plant germplasm does not require reverse engineering to reproduce the germplasm—merely replication. This fact alone is unlikely to alter the conclusion that sale of plant germplasm would constitute disclosure of the trade secret, and trade secret status would therefore be lost. In fact, the ability to replicate the plant germplasm no doubt serves to underline this conclusion, since the trade secret is the germplasm itself.

Thus, the bare sale of a cultivar or hybrid will result in the loss of trade secret protection for the cultivar or hybrid. The issue is slightly more clouded when the sold product is a hybrid, however. The hybrid may contain parental seed, and if it does, the bare sale of the hybrid would be considered to be an accidental disclosure of the parental line(s). That accidental disclosure would therefore result in the loss of trade secret protection for the parental lines.[40] In addition, trade secret protection would not be available if adequate affirmative steps had not been taken to maintain secrecy. Although the court in the *Pioneer* case indicated that Pioneer had indeed taken reasonable precautions, it did not indicate what those precautions were. While the issue was never raised in the Pioneer case, it is possible that Pioneer did not have an effective trade secret policy (which would have substantially weakened its case).

In recent years, restrictive use provisions (further details, infra) have been utilized in connection with sales of plant germplasm. It can be argued that such provisions are a valid means of maintaining trade secret status of a germplasm that would otherwise enter the public domain. If the restrictive use provision is found by the courts to be valid and enforceable, the common practice of obtaining plant germplasm from a farmer's fields may be found to be an unfair means of obtaining a trade secret. However, trade secret protection will still be lost upon independent discovery or reverse engineering, even with the benefit of a restrictive use provision.

CONTRACTS

Contracts are a second alternative to patent and plant cultivar protection, for the protection of plant germplasm. There are several types of contracts that may be utilized, including licenses, sales contracts (also called conditions of sale or purchase agreements), restrictive use labels, bailments, and combinations of these. A common thread between these representations is their attempt to restrict or limit the use of the plant germplasm. The effectiveness of a contract is dependent upon the type of contract. In addition, a contract is only enforceable between the contracting parties, and does not provide any protection against any other parties.

Many of the issues discussed in this section, as they relate to plant germplasm, have not been addressed directly by the courts, although some insight may be gleaned from analogous provisions for other types of information or technology. This section of this chapter will provide a general review of the types of contracts and will discuss their applicability in the protection of plant germplasm.

Licenses

Licenses have wide applicability in a number of areas, including intellectual property (such as patents, plant variety protection certificates, trade secrets, know how, and show how). In fact, patent licenses are a rather common occurrence. In all licenses, the licensee receives the right to enjoy something owned by the licensor without interference from the licensor. This transfer of rights can range from a conveyance of substantially all rights to one of severely restricted rights. The license establishes a continuing business relationship between the parties, but does not establish a fiduciary relationship.[41] A license may contain a confidentiality provision, however. The entire agreement should be examined to determine if it is in fact a license. An agreement may be called a license, an agreement, or a condition of sale, but each could have the same provisions and accomplish the same purpose. Since a license is more substantive than a purchase agreement it generally involves more negotiations between the parties, and it is thus less likely to be held unenforceable when compared to the standardized "boilerplate" language of purchase agreements. Licenses are governed by state laws, a fact that should be kept in mind when drafting enforceable licenses. A license may be of any duration agreed to by the parties, but a license of a patent may not provide for royalties after the expiration of the patent.[42] Royalties can be collected on the license of a patent application even if the application does not mature into a patent. In such a case, there is no limitation on the length of time during which royalties can be collected.[43]

There are three basic types of licenses—express, implied, or license by estoppel or acquiescence. An express license is a written or oral license between a licensor and licensee. An implied license is one that is created by the language or conduct of an intellectual property owner,[44] unless such a license would be contrary to express agreements between the parties. A license by estoppel or acquiescence is created if (i) there is an unauthorized use of the property, (ii) the owner knew of such use, (iii) the conduct of the owner misled the user to believe the owner had abandoned interest or acquiesced in the use, and (iv) there is detrimental reliance by the user (such as the building of a new production facility) on such conduct. In this factual situation, the owner is prevented (estopped) by the courts from asserting infringement of his rights on the basis of the lack of a license.

The grant of a license may be exclusive, nonexclusive, or restrictive. In essence, an exclusive license is an agreement that the licensor will not grant any other licenses after execution of the exclusive license. The exclusive license may be subject to previous nonexclusive licenses, however, and may preclude the licensor from practicing the licensed subject matter unless he has reserved the right to do so. A nonexclusive license is an agreement for the licensee to practice the subject matter, but allows the licensor to grant additional licenses as he/she desires. A restricted license is an agreement in which the right to practice the subject matter is limited. The license can be restricted in numerous areas, but those restrictions most common are by (i) duration, (ii) type of activity or field of use, and (iii) geographic area. An example of a field-of-use restriction is the license of plant germplasm for reproduction purposes only. In this instance, the plant germplasm cannot be lawfully used for breeding, producing hybrids, or the like. Clearly, a restricted license provides a licensor with maximum control of the subject matter.

One important aspect of a license is that the licensee is generally estopped from challenging the validity of licensed rights. Important exceptions to this rule are licenses that include patents.[45] Once a patent has been held to be invalid or unenforceable, no further royalties are due.[46] Since most licenses of patent rights also include know-how rights (i.e., a commercial embodiment of the invention, or additional information that may be necessary to practice the invention), it is important that the license apportion royalty payments between the two, so that royalty payments can still be collected for the license of know-how rights. It is important to address termination provisions in drafting a license, in the event that a patent does not issue or is invalidated, or a trade secret is lost.

As can be determined from this brief discussion, there are a number of factors to bear in mind, beyond the scope of this chapter, when approaching the licensing of intellectual property rights. There are several treatises which discuss these issues in greater detail.[47]

Applicability to Plant Germplasm

Licenses have been used successfully to transfer rights to a licensee of plant germplasm that has been protected by various mechanisms. Examples of licensed plant germplasm patents include those to: (i) a leafy trait, and (ii) herbicide resistance. Corn inbreds protected by PVP certificates have similarly been licensed, although the license may not have referred to such certificates per se. In light of case law, it is clear that plant germplasm can be licensed, regardless of the type of property right. It is equally clear, however, that the type of property right will largely dictate the terms that may be available in the license and the scope of that license. It is more likely that a license under a patent or a PVP certificate will be more substantive than one of a trade secret or other proprietary property right. The major reason for this is that a properly negotiated trade secret license will terminate upon the loss of trade secret protection (which is likely to occur with a plant germplasm trade secret).

A license containing a restricted field of use will be more easily enforceable than a restrictive use provision (discussed further below). However, it is doubtful that a restriction on the use of plant germplasm for breeding purposes will be enforceable if the germplasm is protected by a PVP certificate. The PVPA could likely be held to preempt any such field-of-use restriction, since it would not appear to comply with Congress' intent, in passing the PVPA, that protected cultivars be used for developing new cultivars. This view was also expressed by a USDA administrator in 1980 during hearings on amendment of the PVPA.[48] In fact, seed deposited in connection with a PVP application may be available to the public.[49]

It is worth remembering that the strength of a property right that attaches to a germplasm determines the strength of its licensing program. If a property owner can totally exclude all others from the germplasm, then a license is an excellent vehicle with which to transfer property rights yet retain maximum control. This situation would be more likely to occur if the plant germplasm were protected by a patent that prohibited the manufacture, use, or sale of the claimed plant germplasm. However, little or no control is available if the subject matter could easily slip into the public domain and become accessible to others, which may be the case for trade secrets or other property rights. In that instance, a license could be easily terminated. The value of a license is only as good as the underlying protection for the plant germplasm.

Restrictive Use Provisions

Restrictive use provisions, as designated in this chapter, have been utilized in purchase agreements, on drawings, and on labels in an attempt to limit a purchaser's rights to the purchased material. A restrictive use provision could be viewed analogously to a field-of-use restriction in a license agreement. However, a purchase agreement or a label generally is a nonnegotiable item, and a license agreement is often the result of complex and/or lengthy negotiations. As a result, there are more significant questions concerning the validity and enforceability of restrictive use provisions than with field-of-use restrictions in a license agreement. One reason for this is that a license does not represent the sale of licensed matter, as is the case with a purchase agreement or a label. Restrictive use provisions have generally been held to be enforceable if the purchaser was aware of the restriction.[50]

It is worth noting that many examples of restrictive use provisions have been upheld in the context of trade secret law. For example, in the *Data General* case,[51] Data General provided computer design drawings to computer purchasers for the purpose of enabling them to perform maintenance on the equipment. The drawings bore the legend, "this drawing and specifications, herein, are the property of Data General Corporation and shall not be reproduced or copied or used in whole or in part as the basis for manufacture or sale of the items without written permission." The sale of the computer was accompanied by a standard-form contract that included a confidentiality agreement limiting the use of the drawings to maintenance only. The evidence showed that the defendant's president was aware of the purpose and meaning of the legend on the drawings. His company's copying of the drawing in order to manufacture a similar computer was held to be an improper acquisition of trade secrets.[52]

Restrictive use provisions could be incorporated into a sales contract if they were presented to the buyer. For example, in *Belanger,*[53] the court held that catalog provisions or conditions of sale on a price list could be incorporated into a contract of sale if they were clearly presented to the prospective customer.

The court in *Coast Trading*[54] held that a purchase order must be read in light of the quotation that inspired its submission to the purchaser. Standard terms and conditions found in a quotation to a prospective customer were incorporated into the purchase order.

Restrictive use provisions have also been attempted in drafting computer software licenses, to prevent copying the software or decompiling, disassembly, and reverse engineering of the software. Such provisions have been held to be unenforceable on the basis of preemption by federal copyright statutes.[55]

The use of a restrictive use provision on a label or in a purchase agreement could be considered an attempt to maintain the trade secret status of the restricted materials by creating a contractual or confidential relationship between the seller and the purchaser. The primary question, though, is whether such a label restriction does create a contract with all purchasers. The *Data General* case and related cases would imply that such a legend might be sufficient to extend trade secret protection. These cases can be distinguished on the basis of limited distributions of materials, as opposed to the wide distribution attendant with a large number of sales (e.g., with plant germplasm). If there is a large number of sales, and if a restrictive use provision is successful in extending trade secret protection, then the provision may serve to put any party on notice of the trade secrets, and could then provide a basis for arguing that any forbidden use of the material would be a misappropriation or an appropriation by unfair (albeit legal) means.

Applicability to Plant Germplasm

Two good examples of restrictive use provisions that have been employed for plant germplasm are those found on labels of hybrid corn sold by Pioneer Hi-Bred International (Des Moines, IA) and on purchase agreements for soybean [*Glycine max* (L.) Merr.] cultivars sold by Midwest Oilseeds (Adele, IA). The front of the label contains a notice referring to what Pioneer terms "important information" on its reverse side, which states that:

> The following provisions are part of the terms of sale of this product:
>
> One or more of the parental lines used in this hybrid are the exclusive property of Pioneer Hi-Bred International, Inc. Buyer intends to purchase and seller intends to sell only hybrid seed. Buyer agrees that purchase of this bag of seed does not give any rights to use any such parental line seed which may be found herein,

> or any plant, pollen or seed produced from such parental line seed for breeding, research or seed production purposes, or for any purpose other than production of forage or grain for feeding or processing.

The following caveat is made on the first page of a Midwest Oilseeds purchase agreement, immediately above the customer's signature line: "Terms on the reverse side are a part of this Agreement. Read carefully." A pertinent paragraph of the agreement states as follows:

> In consideration of the foregoing and in consideration of the Midwest Oilseeds Soybean Seeds herein sold, Purchaser hereby acknowledges and agrees that the production from the Midwest Oilseeds Soybean Seeds herein sold will be used only for feed or processing and will not be used or sold for seed, breeding or any variety improvement purposes. Purchaser acknowledges Supplier's proprietary interest in the use of subsequent production from the seeds herein sold, and agrees it would be a violation of this agreement to allow the subsequent production of the seed herein sold to be used to create a seed variety or seed product from said production, which may be used for seed purposes by individuals or entities other than Midwest Oilseeds, Inc. Purchaser agrees and acknowledges that any use of Midwest Oilseeds Soybean Seeds, which is forbidden by this agreement, will constitute a misappropriation of the personal property of Midwest Oilseeds, Inc., and will therefore result in a breach of this agreement. Purchaser agrees that Supplier may bring an action in Dallas County, Iowa, to recover damages as a result of the breach of this agreement, along with reasonable attorney fees and costs associated with any action commenced in regard thereto.

It is interesting to note that the terms in the purchase agreement indicate a minimum damage of $2.84 per 10 L ($10.00 per bu) for a violation of the restrictive use provision. In addition, each seed bag label states that the "use restriction of Seed Purchase Agreement on each order blank" is a "part of the terms of sale."

Although Pioneer has used restrictive use labels on its hybrid seed, it recently alleged in suit that the restrictive use language of Midwest Oilseeds is "invalid, unenforceable, and without legal effect as alleged."[56] If the suit should go to trial, a holding on the enforceability of restrictive use provisions may be rendered. In the interim, however, there may be considerable risk in utilizing the restricted plant germplasm for a restricted purpose, since the provision might involve the application of trade secret law. This risk would attend both the plant germplasm purchaser, and any competitor who was aware of the application of restrictive use provisions and desired to obtain the plant germplasm from the purchaser's field.

Restrictive use provisions, such as prohibiting use for breeding purposes, would be unnecessary for patented plant germplasm, since a license could be given that would restrict the field of use. Such a license could be in the form of a "label license" attached to bags of the patented plant germplasm. Such a license would be enforceable under the patent laws, and would not run the risk of being held unenforceable or invalid, as might be the case for restrictive use provisions.

It is likely that a provision restricting the use of plant germplasm protected by a PVP certificate for breeding purposes would be held to be unenforceable as being preempted by the PVPA, a consequence of the intent of Congress and the language of the act that permits the use of protected germplasm in the development of new cultivars. In fact, there is a strong possibility that any competitor could obtain samples of seed deposited in accordance with the regulations of the PVPA.

Bailment

A third type of contract that may be an alternative for the protection of plant germplasm is a bailment agreement. One definition of *bailment* is the deposit of personal property of one person (the bailor) with another (the bailee), usually with a particular purpose and under some kind of arrangement. In a bailment, the title of the property remains with the bailor, but the bailee is entitled to possession of that property. A classic example of bailment is the parking of one's automobile at a parking garage. If the property is not returned to the bailor or disposed of as provided in the agreement, the bailor can be held to have converted the property. The presence of a bailment should strengthen a claim for conversion of the property.

A bailment has been created when the personal property has been delivered to the bailee. The delivery may be (i) an actual delivery in which the bailee or his agent is given real possession of the property, or (ii) a constructive delivery in which certain acts are held to be equivalent to an actual delivery. A bailment is often created for the execution of a special object upon, or in relation to, the property. The bailment may benefit the bailor, the bailee, or both. Many bailments are beneficial to both parties who generally contemplate some price or compensation in return for the benefits flowing from the bailment and as usually set forth in an agreement. The agreement may be express or implied, and may be written or oral. The agreement will provide for the return of the property or for otherwise disposing of the property in conformity with the purpose of the bailment.

Applicability to Plant Germplasm

Bailments have been proposed as one means of providing protection to the owner of newly developed organisms. This theory was first proposed for microorganisms, hybridomas, or cell lines. Recently, however, bailment has been proposed as an alternative to licensing patented transgenic animals. As previously discussed, the plant genetic message or plant germplasm is intangible personal property. As such, plant germplasm can be the proper subject of a bailment.

A bailment should be established with all parties who will receive or have access to plant germplasm, and should be by contract. The bailment agreement should provide that the owner of the plant germplasm is loaning the plant germplasm (in whatever form it may be or take) to a third party. The agreement should further specify what activities the third party is permitted to do to, for, or with the germplasm. The agreement should specifically state that any progeny of the plant germplasm belongs to the owner of the original plant germplasm. The progeny could be limited to the actual breeding material, e.g., an inbred (for plants in which a hybrid is sold) or the cultivar itself (for plants in which the cultivar is sold). The disposition or return of the plant material or progeny should be detailed in the agreement, as well as the consequences of the disposition. For example, the agreement may provide that a third party bailee may use the progeny for food or sell it as a commodity crop and keep the profits. The agreement should specify that the plant germplasm or its progeny may not be used for breeding unless breeding is the purpose of the bailment. The agreement could allow the sale or reuse of breeding material by a third party with specified compensation to the owner of the plant germplasm.

If a bailee violates the bailment agreement, the plant germplasm owner should be able to recover damages under the tort theory of conversion discussed later. It appears that a valid bailment would enhance the ability of an owner to utilize the conversion theory of law and should avoid some of the issues raised by the court in the *Pioneer* case.

It is likely that a bailment that essentially prevented the development of new cultivars would be held to be unenforceable as being preempted by the PVPA, if the plant germplasm were the subject of a PVP certificate (for reasons previously discussed).

LAW OF CONVERSION

A third alternative for the protection of plant germplasm is under the tort theory of conversion. One definition of *conversion* is the "unauthorized and wrongful exercise of dominion and control over another's personal property, to exclusion of or inconsistent with rights of owner."[57] Although conversion originally applied only to tangible personal property, the term has evolved to be applied also to intangible personal property such as rights under a patent,[58] a computer program,[59] a genetic message of plant germplasm,[60] or one's own genetic message.[61] A rationale for holding intangible property subject to conversion is set forth in *Pearson v. Dodd,*[62] wherein the court stated that:

> Where information is gathered and arranged at some cost and sold as a commodity on the market, it is properly protected as property. Where ideas are formulated with labor and inventive genius, as in the case of literary works or scientific researches, they are protected. Where they constitute instruments of fair and effective commercial competition, those who develop them may gather their fruits under the protection of law.

As the court stated in *United States v. May,*[63]

> The touchstone of conversion is the exercise of such control over property that serious interference with the rights of the owner . . .

The exercise of control that in fact is inconsistent with an owner's rights does not require a matter of conscious wrongdoing,[64] nor is a mistake of law or fact a viable defense. For example, good faith on the part of the alleged converter does not excuse the act from being a conversion.[65]

In order to be held liable under the law of conversion, a party must wrongfully acquire possession of a rightful owner's property and interfere with control of that property. Acquisition may be without legal justification, by fraud or by the unauthorized transfer or disposal of possession of the property to one who is not entitled to it. An example of the latter may include the transfer of possession to one of the original property owner's competitors. In addition, alteration of the property may itself be grounds for a claim of conversion. Alteration prior to sale does not preclude liability for a claim of conversion,[66] nor do modification or enhancement in the value of the property.[67] Additional proof of conversion necessary is that the owner was in possession of the property or entitled to possession of the property at the time of the conversion. A restrictive use provision, as above, may be useful in establishing entitlement to possession. The conversion of property is considered complete as soon as the party acquires, detains, or disposes of the property.

Applicability to Plant Germplasm

There is now some legal precedent to argue that plant germplasm is property that can be converted.[68] The *Moore* case is now returning to a lower court to determine whether the patient's germplasm or cells are property and have been converted. *Moore* could probably be distinguished on its facts, however, since the property (spleen tissue) was not generally available (such as by sales, for example), and was taken from the patient during surgery. In the *Pioneer* case, the court concluded that trade secret law was more applicable to the facts concerning plant germplasm than the law of conversion, and as such, the plant germplasm was not the kind of property protected by the law of conversion. In this regard, the court stated:

> Because of concepts which have been developed as a part of the law of trade secrets, such as independent development of the same "secret" and protection only for secrets which are indeed secret, the Court concludes that the possibility of setting unsound precedent is minimized by ruling that the genetic message of H3H and H43SZ7 is a trade secret and property for the purpose of the law of trade secrets, but is not the kind of property protected for the purposes of the law of conversion.[69]

However, the court concluded that if the genetic message was not ultimately protected by trade secret law, then

> . . . the Court rules as a matter of law that Pioneer's interest in the genetic message is "property" which is protected by the law of conversion. Furthermore, the court finds that Holden asserted wrongful control over the genetic message of H3H and that such control detracted from Pioneer's ability to control how its property was used.[70]

It appears that the judge in *Pioneer* was concerned with providing the plant germplasm owner greater rights under the law of conversion than were available under trade secret law, if the germplasm is proper subject matter for trade secret protection. Greater rights would indeed exist under the law of conversion, since reverse engineering, independent development, and disclosure are not exceptions to liability for conversion. This fact would appear to be a strong argument for the inapplicability of the law of conversion in the protection of plant germplasm. However, in the absence of a final decision concerning these issues, there is some risk that the law of conversion could be utilized to protect plant germplasm. If it can, then obtaining the plant germplasm by purchase or in a purchaser's field would be sufficient acts to constitute conversion, a position that would be strengthened with the use of a valid and enforceable restrictive use provision.

CONCLUSIONS

The applicability of any alternatives to patent and plant variety protection for plant germplasm is directly related to the fact that the germplasm of interest is self-replicating. Alternative means of protection are generally not as valuable as patent or plant variety protection, and certainly patent protection is more valuable and protective than plant variety protection. Trade secret law is not a viable alternative, since sale of a cultivar or a hybrid containing an inbred parent could constitute either a disclosure or an accidental disclosure, with the attendant loss of trade secret protection. Trade secret law would be more viable if restrictive use provisions were found to be valid and enforceable, thus enlarging trade secret protection. In any event, trade secret law does not prohibit independent discovery or reverse engineering. In light of recent biotechnology breakthroughs, reverse engineering is close to becoming a reality, especially in the production of cultivars having substantially identical, essential characteristics.

Licenses of nonpatented or nonplant cultivar-protected plant germplasm are not a significantly more viable alternative to trade secret protection, since there is a risk of loss of control of the material, either by a licensee giving away the plant germplasm or by a third party obtaining the germplasm in a licensee's field. On the other hand, licenses of patented or plant cultivar-protected germplasm are very valuable. A patent license is even more valuable than a PVP certificate license, since a label license can contain a restricted field of use.

Restrictive use provisions on labels or purchase agreements can be quite useful to the germplasm owner, if they are upheld as being valid and enforceable. Such provisions would be most useful if they could extend trade secret protection to cover plant germplasm upon its sale, which would significantly reduce the risk of a disclosure (accidental or intentional) of the trade secret. Although there are good arguments on both sides of the issues of validity and enforceability, there is a considerable risk in utilizing plant germplasm that contains such a restrictive use provision for a prohibited use, regardless of whether the user is a purchaser or a third party.

Bailment, through the use of a bailment agreement, could also be an alternative to patent and plant variety protection. It appears that there are not as many issues relating to the applicability of bailment agreements to plant germplasm as with some of the other previously mentioned alternatives. The remedy for violation of a bailment lies in the law of conversion. A valid bailment agreement should eliminate many of the issues concerning the applicability of conversion to the protection of plant germplasm merely on the basis of the property right.

The law of conversion could also be an alternative to patent and plant variety protection, if plant germplasm were found to be the kind of property protected from conversion. A major issue concerning applicability of the law of conversion centers on trade secrets and the legally proper means of independent development, disclosure, and reverse engineering for discovery of a trade secret (if a trade secret is discovered by these means, there is no liability under trade secret law). It is clear that the law of conversion would provide a greater property right than trade secret law.

Although there are presently a number of alternatives for the protection of plant germplasm, there are many questions concerning the validity, enforceability, and viability of several of these alternatives. These questions will undoubtedly be answered through litigation, and at that time more definitive conclusions can be made as to applicability of these alternatives in the protection of plant germplasm.

FOOTNOTES

1. 35 U.S.C. § 101. A utility patent is directed to processes, machines, manufactures or compositions of matter or improvements therein. Included in this list are plants, plant parts, genes, gene combinations, proteins and the like.
2. 35 U.S.C. § 161. The term "plant patent" is directed to plants which are asexually reproduced under the Plant Patent Act (PPA).
3. 7 U.S.C. §2402. A PVP certificate is directed to sexually reproduced plants.
4. *Diamond, Commr. Pats. v. Chakrabarty,* 447 U.S. 303, 308-9 (1980). S. Rep. No. 1979, 82d Cong., 2d Sess. 5 (1952); H.R. Rep. No. 1923, 82d Cong., 2d Sess. 6 (1952).
5. 227 U.S.P.Q. 443 (BPAI 1985).
6. 15 U.S.C. 1125(a).
7. *Pioneer Hi-Bred International, Inc. v. Holden's Foundation Seeds, Inc.* ____ F. Supp. ____ (S.D. IA. 1987), slip opinion dated October 30, 1987.
8. *Ibid,* at p. 69–70 of the slip opinion.
9. *Sinclair v. Aquarius Electronics, Inc.,* 42 Cal. App. 3d 216 (1974).
10. *Kewanee Oil Co. v. Bicron Co.,* 416 U.S. 470 (1974).
11. *Kewanee Oil, supra.*
12. The actual number depends on which commentator has analyzed the pertinent state statutes.
13. For example, the UTSA would protect disabling or negative information which is not known, but would be of value if known by competitors, as well as ideas or concepts. This information may not be protected under the *Restatement* definition as a result of the use requirement.
14. The subject matter of a patent must be either a process, machine, manufacture, composition or improvement of the same. The subject matter of a PVP certificate must be a sexually reproduced plant variety.
15. Comment b of the *Restatement* lists six factors as relevant to the determination of whether information is a trade secret. These factors are: (1) The extent to which the information is known outside of [the employer's] business; (2) the extent to which it is known by employees and others involved in his business; (3) the extent of measures taken by [the employer] to guard the secrecy of the information; (4) the value of the information to [the employer] and his competitors; (5) the amount of effort or money expended by [the employer] in developing the information; (6) the ease or difficulty with which the information could be properly acquired or duplicated by others. See, *In re Innovative Construction Systems, Inc.,* 793 F.2d 875, 230 U.S.P.Q. 94, 97 (7th Cir. 1986).

16. *Pioneer, supra.*
17. *Ibid,* at p. 72 of the slip opinion. It is interesting to note that in his decision, the judge never specifically identifies the affirmative steps taken by Pioneer to protect the trade secret, which is necessary for establishing a trade secret.
18. See, e.g., *Affiliated Hosp. Prods. Inc. v. Baldwin,* 373 N.E. 2d 1000 (Ill. App. Ct. 1978).
19. See, e.g., *Syntex Ophthalmics, Inc. v. Tsuetaki,* 702 F.2d 677 (7th Cir. 1983).
20. See, e.g., *Financial Programs, Inc. v. Falcon Fin. Servs., Inc.* 371 F. Supp. 770 (D. Ore. 1974); *General Aniline & Film Corp. v. Frantz,* 272 N.Y.S. 2d 600 (N.Y. Sup. Ct. 1966).
21. See, e.g., *National Starch Prods., Inc. v. Polymer Indus., Inc.,* 79 N.Y.S. 2d 357 (N.Y. App. Div. 1948).
22. See, e.g., *Kewanee Oil, supra; National Starch, supra.*
23. See, e.g., *Boeing Company v. Sierracin Crop.,* 108 Wash. 2d 38, 4 U.S.P.Q. 2d 1417 (Wash. 1987); *Electro-Craft Corp. v. Controlled Motion, Inc.,* 332 N.W. 2d 890, 220 U.S.P.Q. 811 (Minn. 1983); *Weil-McLain Co., Inc. v. Andro Corp.,* 200 U.S.P.Q. 538 (N.D. N.Y. 1977); *Clark v. Bunker,* 453 F.2d 1006 (4th Cir. 1972).
24. See, e.g., *Electro-Craft, supra.*
25. See, e.g., *Electro-Craft, supra; Cadillac Gage Co. v. Verne Engineering Corp.,* 203 U.S.P.Q. 473 (Mich. Cir. Ct. 1978).
26. See, e.g., *K-2 Ski Co. v. Head Ski Co., Inc.,* 506 F.2d 471 (9th Cir. 1974); *Sigma Chemical Co. v. Harris,* 605 F. Supp. 1253 (E.D. Mo. 1985), *mod.* 794 F.2d 371 (8th Cir. 1986).
27. See, e.g., *CPG Prod. Corp. v. Mego Crop.,* 214 U.S.P.Q. 206 (S.D. Ohio 1981); *Surgideu Corp. v. Aye Technolog. Inc.,* 648 F.Supp. 661 (D. Minn. 1986), *aff'd.* 828 F.2d 452 (8th Cir. 1987).
28. See, e.g., *Wilson Certified Foods, Inc. v. Fairburg Food Prods. Inc.,* 370 F.Supp. 1081 (D. Neb. 1974); *Valco Cincinnati Inc. v. N&D Machinery Services, Inc.,* 492 N.E. 2d 814 (Ohio Sup. Ct. 1986).
29. See, e.g., *Electro-Craft Corp., supra; In re Innovative Construction Sys. Inc., supra; Digital Development Corp. v. International Memory Sys.,* 185 U.S.P.Q. 136 (S.D. Col. 1973).
30. For citation of relevant cases, see Epstein, M.A. and Neelman, B.K. in *Fourth Annual Biotechnology Law Institute,* Prentice-Hall Law & Business, Clifton, N.J., at p. 220–225 (1988) and Day, L.R., *ibid.* at p. 250–253.
31. *Kewanee, supra,* at p. 475.
32. *Restatement,* Section 759(c); UTSA, Section 1 (1).
33. See, e.g., *E.I. du Pont de Nemours & Co. v. Christopher,* 431 F.2d 1012 (5th Cir. 1970), *cert. denied,* 400 U.S. 1024 (1971).
34. See, e.g., *Curtiss-Wright Corp. v. Edel-Brown Tool Co., Inc.,* 407 N.E. 2d 319 (Mass. 1980).
35. *Restatement,* Section 758; *Curtiss-Wright, supra.*
36. *Restatement,* Section 758.
37. See, e.g., *Garter-Bare Co. v. Munsingwear, Inc.,* 723 F.2d 707 (9th Cir. 1984).
38. *Sinclair v. Aquarius Electronics, Inc.,* 42 Cal. App. 3d 216 (1974).
39. For example, protection may be available in New York [*Williams Press, Inc. v. Flavin,* 44 App. Div. 21634, 353 N.Y.S. 2d 970 (1974)] but not in Michigan [(*Kubik, Inc. v. Hull,* 56 Mich App. 335, 224 N.W. 2d 80 (1974)].
40. The judge in *Pioneer, supra,* indicated that if an inbred was obtained from a farmer's field in which the hybrid was grown, then the inbred may have been obtained by proper means. This issue was not squarely addressed, however, since Holden's did not present it as part of its case.
41. See, e.g., *Laughlin Filter Corp. v. Bird Machine Corp.,* 319 Mass. 387, 65 N.E. 2d 545, 69 U.S.P.Q. 23 (1946).
42. See, *Burlotte v. Thys. Co.,* 379 U.S. 29, 143 U.S.P.Q. 694 (1964).
43. See, *Aronson v. Quick Point Pencil Co.,* 440 U.S. 257 (1979).
44. See, e.g., *DeForest Radio Telephone & Telegraph Co. v. United States,* 273 U.S. 236 (1927).
45. *Lear v. Adkins,* 395 U.S. 653 (1969).
46. *Troxel Mfg. Co. v. Schwinn Bicycle Co.,* 489 F.2d 968 (6th Cir. 1973), *cert. denied* 416 U.S. 939 (1974).
47. See, e.g., *Eckstrom's Licensing in Foreign and Domestic Operations,* Clark Boardman Co. Ltd., New York (1987).

48. U.S. Congress, Senate Committe on Agriculture, Nutrition and Forestry, *Plant Variety Protection Act,* hearings before the Subcommittee on Agricultural Research and General Legislation, June 17–18, 1980, p. 71 (Washington, D.C., U.S. Government Printing Office, 1980).
49. *Ibid.*
50. See, e.g., *Data General Corp. v. Digital Computer Controls, Inc.,* 357 A.2d 105 (Ct. Chan. Del. 1975); *A. Belanger and Sons, Inc. v. United States,* 275 F.2d 372 (1st Cir. 1960); *Coast Trading Co. v. Parma, Inc.,* 587 P.2d 1071 (Ct. App. Wash. 1978).
51. *Data General, supra.*
52. See also, *Rexnord, Inc. v. Ferris,* 637 P.2d 619 (Ore. 1981); *J&K Computer Systems, Inc. v. Parrish,* 642 P.2d 732 (Utah 1982); *Digital Development Corp. v. International Memory Systems,* 185 U.S.P.Q. 136 (S.D. Cal. 1973).
53. *Belanger, supra.*
54. *Coast Trading, supra.*
55. *Vault Corp. v. Quaid,* 655 F.Supp. 750 2 U.S.P.Q. 2d 1407 (E.D. La. 1987).
56. See paragraph 21 of Pioneer's answer in *Midwest Oilseeds, Inc. v. Pioneer Hi-Bred International, Inc.,* Case No. 88-1328A (S.D. Iowa).
57. *Black's Law Dictionary,* 5th ed., West Publishing Co., St. Paul, Minn. (1979). See also, e.g., *Wilke v. The City of Davenport,* 309 N.W. 2d 450 (Iowa 1981); *Restatement (Second) of Torts,* Section 242(2).
58. See *Miracle Boot Puller Co., Ltd. v. Plastray Corp.,* 225 N.W. 2d 800 (Ct. App. Mich. 1975).
59. See *National Surety Corp. v. Applied Systems, Inc.,* 418 So. 2d 817 (Sup. Ct. Ala. 1982).
60. *Pioneer, supra.*
61. See *Moore v. The Regents of the University of California et al.,* 202 Cal. App. 3d 1230, 249 Cal. Rptr. 494 (Ct. App. Cal. 1988).
62. *Pearson v. Dodd,* 410 F.2d 701 (D.C. Cir.), *cert. denied* 345 U.S. 947 (1969).
63. 625 F.2d 186 (8th Cir. 1980).
64. See, e.g., *Allred v. Hinkley,* 328 F.2d 726 (Utah 1958).
65. See, e.g., *Poggi v. Scott,* 139 P. 815 (Cal. 1914); *McGlynn v. Schultz,* 218 A.2d 408 (Supp. Ct. N.J. 1966), *aff'd.* 231 A.2d 386 (App. Div. N.J. 1967).
66. *Pioneer, supra.*
67. *Moore, supra.*
68. *Pioneer, supra; Moore, supra.*
69. *Pioneer, supra,* at p. 81 of slip opinion.
70. *Ibid.,* at p. 81–82.

5 Issues and Challenges in the Administration of the Patent Law with Regard to Plants by the Patent and Trademark Office

Charles F. Warren
U.S. Patent and Trademark Office
Washington, DC

The current Patent Law, including the Plant Patent Act, confer upon the Patent and Trademark Office (PTO) the authority to administer the granting of patents on plant subject matter. Patents can be granted under either the general (utility) statute (35 U.S.C. 101) or under the plant patent statute (35 U.S.C. 161).

The administrative issues and challenges encountered in the granting of such patents arise primarily in the area of examining practice and procedure, although some purely administrative matters, such as staffing requirements, are also present. The resolution of these matters by the PTO is ongoing and evolutionary, with thorough consideration of the input from a varied constituency. Generally, these issues and challenges are similar to others encountered in the biotechnology arts. However, some are unique to the examiniation of applications in the plant arts. The principal area generating issues and challenges is in the examiniation of plant utility applications. In some instances, the same matters also affect the examination of plant patent applications.

BACKGROUND: UNIT ORGANIZATION, STAFFING, AND CASELOAD

Patent Examining Group 180 (Biotechnology) has the responsibility for examination of all plant utility and plant patent applications directed to plant biological materials, including plants per se, as well as products and processes relating thereto. This Group was formed on 28 Mar. 1988, in response to the emergence of biotechnology as an art area and the growing backlog of applications related to biotechnology. Patent applications for plant subject matter are examined in Art Unit 184. This Art Unit was organized with two

Copyright © 1989 Crop Science Society of America, American Society of Agronomy, Soil Science Society of America, 677 S. Segoe Rd., Madison, WI 53711, USA. *Intellectual Property Rights Associated with Plants*, ASA Special Publication no. 52.

senior examiners along with Class Plants, directed to subject matter under 35 U.S.C. 161, and selected culture and plant process subclasses from utility Class 47, Plant Husbandry, from Group 330, together with several examiners from Group 120 working with the plant, genetic engineering, and cell and tissue culture subject matter provided for in utility Classes 435, Chemistry: Molecular Biology and Microbiology, and 800, Multicellular Living Organisms and Unmodified Parts Thereof.

The current staffing for the plant art areas includes two senior primary examiners, Mr. Bagwill and Mr. Feyrer, who examine applications in Classes Plants, 47 and 800, and five junior examiners, Mr. Fox, Mr. Cohen, Ms. Rhodes, Mr. Benzion, and Ms. Chereskin, the latter three newly hired, who primarily examine applications in Classes 47, 435, and 800, but will also be exposed to applications filed under 35 U.S.C. 161. The junior examiners are technically qualified in the range of plant arts examined. They are becoming acquainted with examining practice and procedure through the Patent Academy of the Patent and Trademark Office and are participating in law courses offered by the Academy. The Patent Academy is also training all seven examiners in the use of various commercial data bases relevant to the art area.

To keep examiners abreast of the latest technical and legal developments in the biotechnology art area, the PTO with the cooperation and assistance of private trade associations is developing the Biotechnology Institute. One of the purposes of the Biotechnology Institute is to provide supplemental training for biotechnology examiners in the technical and legal aspects of examining biotechnology inventions. One facet of this training will be directed to the plant art areas and tailored to meet the needs of both senior and junior examiners. Under current plans, the Institute will be a cooperative effort of government and bar and trade associations to provide examiners and others in the PTO (e.g., Quality review Examiners, Examiners-in-Chief of the Board of Patent Appeals and Interferences) with the necessary resources to keep current on law and technology. Public and private sector organizations will be invited to provide the technical and legal resources needed to design and present training courses, and provide the PTO with general reference materials. The Institute will have a governing board of public and private sector representatives that will assist the PTO in the identification of Institute needs and provide resources for the program implementation of the Institute. The major PTO participants include the Patent Academy, which will be responsible for coordinating the diverse activities of the Institute; the Scientific Library, which will acquire literature relevant to the biotechnology examiners' needs; and, Group 180, which, of course, has the examiners to be trained and will provide speakers/educators in specific fields of biotechnology and the law.

The plant utility applications and the plant patent applications are considered to be separate elements of the caseload in the plant art area, which reflects the difference in content and examination requirements between these applications. The filing rate of the plant utility applications has been increasing at a dramatic rate compared to that of the plant patent applications. The filing rate for applications in the plant utility classes has substantially increased from a rate of 73 applications in Fiscal Year 1986, to 137 applications in Fiscal Year 1987, and to 235 applications in Fiscal Year 1988, which are rate increases of 88 and 72%, respectively. Preliminary indications are that the filing rate for Fiscal Year 1989 will be in the 400 case range, which is roughly the same percentage increase as experienced in Fiscal Year 1988 the filing rate. As of 1 Jan. 1989, the inventory of plant utility applications was 211 with about 17 months pendency to first action.

The filing rate for the established area of plant patent applications was at a level of 425 cases for Fiscal Year 1987 that was a significant increase over the filing rate of 342 cases for Fiscal Year 1986, an increase of 24%. The filing rate for Fiscal Year 1988 was 399 cases which represented decrease over Fiscal year 1987. Preliminary indications are that the filing rate for Fiscal Year 1989 will again be in the 400 case range, reflecting a relatively stable filing rate. As of 1 Jan. 1989, the inventory of plant patent applications was 196, with about 7 months pendency to first action.

It is anticipated that the expected increase in the filing rate of plant utility applications will be more than offset by an increase in the examining efficiency of the junior examiners, such that pendency to first action should be reduced to 6 months or less by the end of the fiscal year. However, additional highly qualified examiners will be hired if filing rates so dictate.

STATUTORY PROTECTION, TYPES OF APPLICATIONS, AND FILING REQUIREMENTS

The scope of plant subject matter for which a plant utility patent or a plant patent may be granted differ extensively. Under 35 U.S.C. 161, a plant patent may be granted only for a single new plant variety capable of asexual reproduction, even though the variety also may be capable of stable sexual reproduction and may even be the result of an invention or discovery involving genetic engineering. Plant patents cannot be granted to a tuber propagated plant or a plant found in an uncultivated state. The plant subject matter that may be patented under 35 U.S.C. 101 encompasses plants per se, generically as well as varieties, and plant parts, cell culture, and other plant products and processes that are the result of human ingenuity. As such, the general utility statute provides for patent protection of any asexually or sexually reproducible varieties. This includes varieties that are protectable under the Plant Patent Act (PPA) or the Plant Variety Protection Act (PVPA) as well as hybridized varieties, and plants and microorganisms that are not protected under either of these Acts.

The apparent overlap of subject matter that can be protected under the different statutes was recognized in *Ex Parte Hibberd*, 227 U.S.P.Q. 443 (BdPatApp&Int, 1985), which held that the PVPA and the PPA are not exclusive forms of protection for plant life that would conflict with the granting of patents thereto under 35 U.S.C. 101. See Notice, "Plant Life—Patentable Subject Matter," 1060 OG 4, 8 Oct. 1985.

While it is apparent that the applicant may choose the statutory protection for a new variety, the issues of whether a new variety may be protected under different statutes, and whether a new variety and plant parts and seeds thereof must be protected under the same statute have not been settled. To date, there has been no judicial determination of whether simultaneous protection rather than alternative protection for plant life under these statutes would be appropriate. See S.D. Schlosser, "Legal Protection for Biotechnology in the United States of America"; *Vth International Colloquium on the Protection of Plant Breeders' Rights—The Protection of Biotechnological Inventions in the Field of New Plant Varieties—Legal Problems and Practical Solutions;* CIOPORA; Geneva, Switzerland. May 1988. p. 13–15.

However, the claimed fruit and propagating material from an asexually reproduced plant contained in a utility application have been held to be patentably indistinct (35 U.S.C. 103) from the plant itself that was claimed in a plant patent application. The claimed fruit was considered to be the natural and inevitable product of the plant, and the claimed propagating material was considered equivalent to the plant itself or any embryonic version thereof. An appropriate terminal disclaimer was accepted as obviating the ground of obviousness-type double patenting.

Applicants have attempted to combine claims to different types of patent statutory protection in one "hybrid" application. However, such an application is improper. An application for patent must be directed to only one form of statutory protection, as utility and plant patent applications are subject to different statutory and rules provisions, e.g., specification requirements under 35 U.S.C. 112 and 162, language of the oath or declaration under 37 C.F.R. 1.63 and 1.162, filing fees and maintenance fees. If applicant files a "hybrid" application in a manner that attempts to combine the filing requirements for both forms of protection, the applicant will be given an opportunity to elect the statute under which examination is to proceed, and to file a preliminary amendment placing the application in the appropriate form for the type of application and examination chosen. A hybrid application filed under one form of statutory protection but containing claims to both types of protection, will be substantively treated under the provisions pertaining to the type of protection under which the application was filed, with appropriate rejections made of the claim or claims to the other type of protection.

For completeness, it is noted that plant utility patents and plant patents may be the subject of reissue or reexamination proceedings under 35 U.S.C. 251 and 35 U.S.C. 302 under the same rules that apply to utility patents.

EXAMINATION PROCESS

General Considerations

Both plant utility and plant patent applications are subject to the same examination process as any other national application. As such, the statutory provisions with regard to utility, novelty, nonobviousness, and the written disclosure and claim specificity requirements clearly apply (35 U.S.C. 101, 102, 103, and 112).

The sole exception in terms of the extent of applicability of these statutory provisions to plant subject matter is set forth in 35 U.S.C. 162 with regard to plant patent applications. In effect, Sec. 162 provides that plant patent applications do not have the same technological disclosure requirements or the same claim specificity requirements under 35 U.S.C. 112, first and second paragraphs, respectively, as utility patent applications whether directed to plant or other technology. Under Sec. 112, first paragraph, the written disclosure of a utility application must describe, enable, and provide the best mode for the claimed invention to the extent that one skilled in the art to which the invention pertains must be able to reasonably understand and use the technology behind the invention. However, Sec. 162 requires only that the description of the variety in the plant application need be as complete as is reasonably possible. Similarly, while Sec. 112, second paragraph, provides that the claim(s) in the utility application must particularly point out and distinctly claim the invention, Sec. 162 requires that the sole claim shall be stated "in formal terms to the plant shown and described" in the plant patent application.

Thus, as a practical matter, under the reduced disclosure and claim requirements of Sec. 162, the written disclosure of a plant patent application must at least be as complete as reasonably possible, and provide at least a botanical description of the claimed variety to the extent that the variety is distinguished over all other related known varieties, and its antecedents, and the sole claim must at least recite the new and distinct variety as described and illustrated. See 37 C.F.R. 1.163 and 1.164; M.P.E.P. 1605.

The plant utility application must fully comply with the requirements of 35 U.S.C. 112, as such an application does not fall within the purview of 35 U.S.C. 162.

It is clear that the statutory requirements of novelty and nonobviousness apply equally to plants per se claimed either in a utility application under 35 U.S.C. 101 or in a plant patent application under 35 U.S.C. 161. The examiner has the burden of proof in establishing the basis for the rejection under either of these statutory requirements. Indeed, the most significant aspect of the examination process is the resolution of the issue of nonobviousness in which the examiner must determine the level of ordinary skill in the art to which the claimed invention pertains in order to determine the patentable significance of the difference between the claimed invention as a whole and the closest prior art at the time the claimed invention was made.

The level of ordinary skill in the arts pertaining to plant breeding can be determined, notwithstanding the decision in *Yoder Brothers, Inc. v. California-Florida Plant Corp.*, 1983 U.S.P.Q. 164, at 291–293 (5th Cir., 1976). The Court of Appeals for the Federal Circuit has described several factors, including knowledge and experience, which must be considered in determining the level of ordinary skill in the art to which the claimed invention pertains. See, generally, *Orthopedic Equipment Company, Inc. v. United*

States, 217 U.S.P.Q. 193, at 198–199 (Fed. Cir., 1983). It is apparent that the Court in *Yoder Brothers* did not apply that level of analysis with regard to the level of ordinary skill in the plant arts in considering the issue of nonobviousness.

The plant breeder of ordinary skill is indeed a multifaceted talent. Note, for example, the statement of the "Qualifications of the Plant Breeder," in the paper presented by Dr. Frederick A. Bliss titled "Plant Breeding, Crop Cultivars and the Nature of Genetic Variability" at the Plant Patenting Workshop held in Anaheim, CA, 1 Feb. 1989. The examiner must consider those qualifications of the plant breeder that are relevant to the art to which the claimed plant pertains, following guidelines such as those in *Orthopedic Equipment Company*, in order to determine the level of ordinary skill in that art which will be the standard applied to resolve the issue of nonobviousness in that particular plant utility or plant patent application.

One of the issues in resolving whether the claimed invention is nonobviousness over the art is the extent to which one of ordinary skill in the art would have predicted the result achieved by applicant, given the teachings of the prior art. Thus, in order to make out a *prima facie* case of obviousness of the claimed invention, the examiner must use prior at patents and scientific literature to establish the level of ordinary skill in the art, as well as show that one of ordinary skill in the art would have expected a reasonable chance of success in obtaining the result achieved by the applicant. The applicant may rebut the examiner's position with evidence of unexpected results establishing that the claimed invention is nonobvious. See *In re* O'Farrell, 7 U.S.P.Q.2d 1973, at 1981 (Fed. Cir., 1988).

With regard to the applicability of a plant patent as a prior art reference, it may be necessary for the examiner to establish the availability of the plant reasonably described and illustrated in that reference vis-a-vis the claimed invention, in order to enable the subject matter of the plant patent for purposes of establishing lack of novelty or obviousness.

Plant Utility Applications under 35 U.S.C. 101

Plant utility applications present issues that do not arise in plant patent applications. The claims of plant utility applications may be directed to a variety of plant biological material and processes as noted above, including, genes, proteins, gene and/or protein fragments, recombinant processes, culturing techniques, seeds, and plants per se, generically and/or variety. Within the last several years, the trend has been to file plant utility applications containing claims directed to a diverse variety of subject matter. The diversity of this subject matter generally indicates that a plurality of inventions are present, in which case, the application is reviewed to determine if restriction may be required between the different inventions. Restriction will be required between the inventions if it is established that more than one independent or patentably distinct invention is present, and there is a burden to the PTO to examine more than one of the inventions in the same application. The rules governing restriction in national and Patent Cooperation Treaty utility applications are applicable to plant utility applications.

For example, depending on the facts of the case, the following inventions would be considered to be patentably distinct: the gene and gene fragments; the plasmids containing the gene or fragment and the microorganism hosts containing the plasmids; the transformed cells and the transformed plants; and, the transformation methods. Patentable distinctness is found between these inventions if each invention can be shown to be separately useful from the other inventions. The burden to the Office is demonstrated not only in the different classification and search required for each of these inventions (note that within the inventions, the plasmids and the hosts containing the same are classified and searched in different areas, as are the transformed cells and plants), but also in that the subject matter is recognized in the art to be divergent areas of inventive endeavor.

The use of nucleic acid and/or protein sequence data to define and claim the invention is increasing in biotechnology applications, including plant utility applications. The variety of formats in which this information is submitted in applications coupled with the increasing amount of sequence information, has had a direct impact on the examining process, especially in searching the prior art. In the near future, the PTO will publish a Notice of Proposed Rulemaking regarding requirements for patent applications containing nucleotide and/or amino acid sequence disclosures. The purpose of these proposed rules is to standardize the format for descriptions of nucleotide and amino acid sequence data submitted as part of patent applications in order to facilitate submission of this data in computer readable form upon filing the application.

The sequences subject to the proposed rules have four or more L-amino acids commonly found in naturally occurring proteins, and 10 or more nucleotides that are represented by the four-letter DNA and RNA codes. The submission of the data in machine readable form will improve examination efficiency in terms of improved sequence search capability and interference detection; improve the accuracy of printed sequences along with creation of a data base of all patent-disclosed sequence data falling within the sequence unit limits; facilitate exchange of published patent sequence data, in electronic form, with the Japanese Patent Office (JPO) and the European Patent Office (EPO); and promote conformity with format usage in the scientific community. It is anticipated that submission of sequence data on diskette will provide a direct tie-in with current computerized sequence search programs as outlined next.

In order to effectively examine the ever increasing sequence data appearing in the specification and claims in the plant arts and other art areas of biotechnology, the Office has installed a network of 10 AST PREMIUM 386 machines and a Sun Microsystems, Inc., SUN-3 workstation to run sequence searching software by IntelliGenetics, Inc. (Mountain View, CA), against nucleic acid and protein sequence data bases, including EMBL, ENZYMES, KeyBank, GenBank, PIR(NBRF), VectorBank, and Swiss-Prot. This system is configured to grow with the ever expanding sequence data base information that, of course, is important prior art. The proposed rules for submission of nucleotide and amino acid sequence information in computer readable form, discussed previously, are expected to provide for access to sequence information in applications and patents in a manner compatable with software such as that provided by IntelliGenetics. Indeed, the in-house system is expected to result in substantial savings over the use of the same software/data bases on the BIONET system.

In plant utility applications, as well as other biotechnology utility applications, a deposit of biological material has been accepted as a means of supplementing the written disclosure when that biological material is required to support the claims under 35 U.S.C. 112, first paragraph. It is not readily available to the public or cannot be obtained by a repeatable method set forth in the specification without undue experimentation. The PTO has published a notice of proposed rule making regarding the Deposit of Biological Materials for Patent Purposes, 53 F.R. 39420, 6 Oct. 1988. The purpose of the proposed rules is to define the procedures and requirements to be followed during prosecution of the application once it is determined that a deposit is required to support one or more claims presented for examination, rather than when or under what circumstances a deposit is required.

However, the response to the comments made with regard to the prior notice on this matter, 52 F.R. 34080, 9 Sept. 1987, does shed some light on Office policy in plant utility applications. With regard to a comment concerning the minimum number of seeds to be deposited to ensure availability thereof through the enforceable life of the patent, it was noted that the proposed rules will not set a minimum number. However, analogous to Department of Agriculture requirements for a grant of a Plant Variety Protection Certificate, the Office will consider 2500 seeds to be a minimum number in a normal case, and the applicant may present justification why a lesser number of seeds would be suitable under the circumstances of a particular case. With due regard for the time required for reproduction of seeds, the Office will require a deposit of that number which is likely to satisfy a reasonable demand for samples. As also noted, whether a deposit is required or what type of deposit would satisfy such a requirement in a case depends on the facts of each individual case. This is especially true with cases claiming a hybrid variety where a variety of options with regard to the biological material to be deposited often occurs. In response to a comment concerning the quarantine and inspection procedures encountered in depositing seeds in the USA, one important point was underscored; where a depository does not exist to accept a biological material that would need to be deposited before a patent was granted, the Office would take the position that the requirements of 35 U.S.C. 112 were not satisfied for such an invention.

Several notes on deposits: The Examiner should require a deposit *only* where the biological material is essential to the claimed invention *and* it is not clear from the record that this material is either readily available or obtainable by a repeatable process set forth in the specification. A deposit may be required for different parts of 35 U.S.C. 112, first paragraph. For example, a deposit may be used to satisfy both the enablement requirement, as well as the best mode requirement. However, if the deposit is not required to satisfy the enablement provision Sec. 112, as the biological material required to support the claim may be obtainable without undue experimentation, it sometimes may be necessary to deposit that biological material to satisfy the best mode provision if that exact same material cannot be reproduced by the procedures set forth in the written disclosure. See *In re Wands*, 858 F2d 731, 8 U.S.P.Q. 2d 1400 (Fed. Cir., 1988). Furthermore, only the deposit of the biological material necessary to render the disclosed process repeatable is considered to satisfy the provisions of the statute. For example, if the process for making a product as recited in the written disclosure is repeatable once the starting biological materials are obtained, then the deposit of the starting biological materials, and not the product of that process, would satisfy the requirement.

Where color is a distinctive feature of the plant claimed in a plant utility application under 35 U.S.C. 101, the color shall be positively identified in the specification by reference to a designated color, as given by a recognized color dictionary or color chart in order to comply with the description requirement of 35 U.S.C. 112, first paragraph. Additionally in such case, a color drawing may be necessary for the understanding of the subject matter sought to be patented. 37 C.F.R. 1.81. Unlike 37 C.F.R. 1.165, which provides for color drawings in plant patent applications, drawings in all utility applications are required to be in black and white under 37 C.F.R. 1.84. Accordingly, where applicant recognizes that the plant invention in a utility application requires a color drawing or the examiner determines that a color drawing is necessary under the circumstances of the case for compliance with 35 U.S.C. 112, the applicant should proceed according to the Notice, "Use of Color Drawings in Utility Patent," 1082 OG 25, 24 July 1987. This procedure involves a petition under 37 C.F.R. 1.183 to waive the requirements of Sec. 1.84 to the extent that color drawings on DIN size A4 sheets will be accepted for examination, copying, and archival purposes, and includes five sets of these color drawings. The petition must also include a proposed amendment to the specification to the effect that the file of the patent contains at least one drawing executed in color, a copy of which will be provided by the Office upon request and payment of the necessary fee. The patent copies at issue will depict the drawing in black and white only, but a set of color drawings will be attached to the Letters Patents. The petition will be granted only upon a determination by the Office that a color drawing is the only practical medium by which to disclose in the printed utility patent the subject matter sought to be patented. If circumstances warrant, it is advisable to file any desired color drawings as part of the original application papers in order to avoid issues under 35 U.S.C. 112, first paragraph (compliance with the written description and/or enablement requirements) and 312 (new matter must be avoided), but filing may be deferred until acceptable drawings are required by the examiner.

Plant Patent Applications under 35 U.S.C. 161

A plant patent application can contain *only* one claim, and that claim can encompass only the new distinct variety of the specified plant as described and illustrated. 35 U.S.C. 162; 37 C.F.R. 1.164; MPEP Sec. 1605; see also *Ex Parte Hibberd*, 227 U.S.P.Q. 443, at 447 (BdPatApp&Int, 1985). As set forth previously, a significant difference between the plant patent application and the plant utility application is that the former must contain a description that is as complete as reasonably possible, which is a lesser burden than the requirement to comply with 35 U.S.C. 112, which applies to plant utility applications. It is noted here that the proposed rules with regard to deposit of biological material does not apply to plant patent applications and the Office does not take the position that a deposit is required under the present provisions of 35 U.S.C. 162, even where a deposit has been made in conjunction with a 35 U.S.C. 101 patent.

Where the examiner considers it necessary to the examination of the plant patent application, a duplicate file and drawing of the application are forwarded to the National Program Leader for Horticultural Crops, Agricultural Research Service (ARS), U.S. Department of Agriculture along with a request for a report as to whether the plant variety disclosed is new and distinct over known plant varieties. The reported opinion of the Plant and Natural Resource Sciences staff is taken into consideration by the examiner in preparing the first action in the application but not made of record in the file of the plant patent application. About 80 reports are requested annually. Generally, the time required for the ARS to prepare and return the reports has been increasing over the past year. In view of this and the reduced pendency to first action, plant patent applications are being reviewed upon their receipt in the examiner's docket and promptly forwarded to the ARS if the examiner considers a report necessary to the consideration of the case.

Issues Common to Both Types of Applications

The search of the prior art relative to plant inventions, as is the case with other biotechnology inventions, has presented some challenges that are currently in the process of being met. Biotechnology examiners, including those in the plant art areas, make extensive use of computerized search systems. A primary search tool for U.S. patents is the PTO's own Text Search and Retreval System, a part of the Automated Patent System (APS), which covers a data base of U.S. Patents from 1975 to present, as well as data base of English language abstracts of Japaneses patent documents from 1980 to the present. The majority of the literature references cited by the biotechnology examiners are obtained from searching commercial data bases, such as DIALOG and CAS, which also cover foreign patent literature.

In addition, to further facilitate examiner efficiency, and at the same time provide a testbed for evaluation of automated searching and other examining services, Project 400 of the APS will provide each examiner in Groups 180 and 220 with his/her own PC. The PCs will provide a variety of examining services, including word processing, but are primarily for access to the APS Text Search and Retrieval System and to commercial data bases. The expected examining benefit is a reduction in a patent pendency through a reduction in the time required for search and creation of an Office action, and the mailing thereof to the applicant's attorney or agent. Delivery of these machines to examiners in Group 180 is expected to be about April 1989.

Further, efforts are being made to identify further sources of prior art that pertain to plant utility and/or plant patent applications and determine/acquire efficient systems to search this prior art. It is anticipated that the Biotechnology Institute (discussed previously) will be most helpful in this regard, since one purpose of the Institute is to develop search tools that will improve an effectiveness and efficiency of the examining process in biotechnology. Most of the literature prior art is now identified from computer searches of commercial data bases. It is apparent from prior art statements and other sources that there is additional prior art known to the applicant and those involved in the preparation of the applications that has not been seen by the examiner in a search utilizing the search tools currently available in the Office. Such a system is the GRIN system and other data bases, such as the data base of corn (*Zea mays* L.) variety descriptions, of the Department of Agriculture.

The increase in examining staff, not only in the plant arts but in other areas of biotechnology, has resulted in an increased work requirement on the Scientific Library of the PTO, with regard to acquiring additional search resources and additional holdings to more quickly respond to the examiners' reference requirements. It is anticipated that the Biotechnology Institute (discussed previously) will be of assistance to the Scientific Library in these endeavors. The Scientific Library has over the last 6 months substantially increased their holdings of biotechnology titles and journals in general and the plant area in particular. In order to more effectively serve the needs of Group 180, the Scientific Library is establishing a satellite called the Biotechnology Information Center in connection with the previously established Biotechnology-Chemical Library. The function of the Center is to provide full-time librarians on-site to assist with sequence and other data base searching, as well as with the reference requirements of biotechnology examiners.

As usual, it is requested that applicants and others with a duty to disclose submit information disclosure statements in plant utility and patent applications as soon after filing as possible for consideration by the examiner in preparing the first Office action. The timeliness of filing the statement will become increasingly important as pendency to first action of plant applications is reduced.

The applicant may file a petition to make the application special for examination purposes in either the plant utility application or in the plant patent application. An application accorded a "special" status will be advanced out of turn for examination and treated in an expedited manner throughout its course of prosecution. In view of the differences in pendency to first action, this strategy is more suited to the plant utility applications as the backlog in the plant patent area is being reduced. In addition to the usual basis for petitions to make special, which applies to all utility applications, an interim procedure to make certain biotechnology applications special in view of the backlog in the biotechnology art area has been announced in the Notice, "Petitioning to Make Applications Relating to Biotechnology Special," 1092 OG 55, 19 July 1988. Under this procedure, a small entity applicant who, on petition, can demonstrate that the claimed invention is a major asset of a small entity (individual, nonprofit, or small business) and that the development of this invention will be significantly impaired if examination is delayed, and pays the petition fee of $72, can have his/her application made special for examination purposes. This procedure will be discontinued as the average pendency approaches 18 months in the biotechnology arts.

Under Articles 6 and 13 of the International Union for the Protection of New Varieties of Plants (UPOV) Convention, each plant variety set forth in a claim, in both plant utility and plant patent applications, must be given a variety denomination and that denomination recorded at least by the time that the patent is granted. The PTO published a notice of proposed rule making regarding Variety Denomination Requirements for Plant Patent Applications, 52 F.R. 42016, 2 Nov. 1987. Comments were received and a hearing held in January 1988. In the near future, the PTO will publish a notice of final rule on the matter. One of the administrative issues that must be resolved involves the recordation of the variety denomination, which while not conveying any legal rights in that denomination, still must be examined to determine if there exists a possible conflicting trademark property right, such as a registered trademark or application for registration. It has been proposed that this examination include an opposition proceeding complete with publication of the denomination for the purpose. Such proceedings under current procedures would materially add to the pendency of any application containing a claim to a specific variety under 35 U.S.C. 101 or 161. Consequently, procedures are being explored that would permit a thorough examination of the proposed denomination in a more limited time frame. It should be noted that this rules package would not apply to any protection sought under the PVPA.

Finally, increased activity in granting both plant utility patents and plant patents has created a need for reclassification of Class 800, Multicellular Living Organisms and Unmodified Parts Thereof, and Class Plants (patents under 35 U.S.C. 161). Class 800 currently contains 62 patents, both original and cross-references, consisting of 61 plant utility patents and the "Harvard Mouse" (has its own food supply). The current plans for reclassification of this subject matter provide for the separation of mammalian and plant subject matter, and, within the plant subject matter, a further broad separation based upon plants per sc (according to the method of origination), and plant parts and seeds thereof. Future plans for reclassification of this area are being considered, with efforts being made to identify emerging areas of inventive endeavor within the plant arts. The reclassification plan in class plants is progressing. About one-half of the subclasses in this class of 6516 plant patents will be affected. The emphasis, to date, has been on ornamental plants with the delimitations based upon genus and/or color, and/or other characteristic depending upon the characteristics which define subgroups within each present subclass.

6 Issues and Challenges in the Administration of the Plant Variety Protection Act

Kenneth H. Evans and Eldon E. Taylor

Plant Variety Protection Office
U.S. Department of Agriculture
Beltsville, Maryland

The Patent Act grants ownership rights on new inventions. The 1930 Townshend-Purnell Act amended the Patent Act to provide patent rights to breeders of asexually reproduced cultivars. This amendment is commonly referred to as the Plant Patent Act (PPA). Sexually reproduced cultivars were not included in the patent amendment because they were not believed to be sufficiently identifiable, uniform, and stable to be protected. However, by 1961 it was evident that sexually reproduced cultivars could be produced with uniform and stable recognizable traits. American seed interests, with this knowledge and a desire to provide rights to breeders of sexually reproduced cultivars, considered possible methods of protection. A proposal to extend the PPA to cover sexually reproduced cultivars was considered, but the proposal was opposed by the Patent and Trademark Office (PTO) and failed.

Having failed at amending the PPA to include sexually reproduced cultivars, the American seed interests proposed a bill based on the PPA and the European breeder's rights laws. The basic principles of the proposed bill were: participation should be voluntary; rights should be based only on novelty; no performance tests should be required; protection should not interfere with germplasm distribution; and enforcement of rights should be the responsibility of the owner.

The Plant Variety Protection Act (PVPA) was passed on 24 Dec. 1970, establishing a Plant Variety Protection Office (PVPO) in the USDA to administer the PVPA. Congress enacted the PVPA to provide protection to sexually reproduced cultivars, since protection was not available under the Patent Act. The law, as passed, also included a safeguard for farmers to save seed for their own use and to sell some to their neighbors. It also provided that the applicant could request that the certificate require that the cultivar be sold by cultivar name, only as a class of certified seed. In addition, the PVPA provides for the Plant Variety Protection Board made up of mem-

Copyright © 1989 Crop Science Society of America, American Society of Agronomy, Soil Science Society of America, 677 S. Segoe Rd., Madison, WI 53711, USA. *Intellectual Property Rights Associated with Plants*, ASA Special Publication no. 52.

bers representing the seed trade, the public sector, and farm interests. The board advises the Secretary of Agriculture on administration of the PVPA and on appeals from examiners.

ADMINISTRATION OF THE PLANT VARIETY PROTECTION ACT

The PVPA requires that protectable cultivars clearly differ by one or more characteristics from all prior cultivars of public knowledge. This requires that an application cultivar be differentiated from all prior cultivars of public knowledge before a certificate can be issued. In order to make these comparisons efficiently and accurately, the PVPO has compiled computerized descriptions of over 32 000 cultivars of 100 different crops, which can be searched to eliminate all cultivars that clearly differ in a major characteristic from the application cultivar. An examiner differentiates the remaining cultivars in writing before recommending certification.

Maintaining and updating these descriptive search files is a problem due to the time required to review relevant literature and add the useful information to the computer files. Examiners must interpret the many different descriptive terms used in the literature and select the appropriate computer code to use. Maintaining current computer files is time consuming, but necessary to assure accurate search verifications.

The literature search, however, is facilitated by the PVPO being located in the National Agricultural Library (NAL). This provides convenient and prompt access to the approximately 2 million volume collection of the library and its related bibliographic and automated retrieval services. The NAL acquires approximately 26 000 journals and 15 000 books and reports annually. The PVPO uses the services of the Current Awareness Literature Service (CALS), a computer-based system designed to keep researchers and others posted on current literature. The CALS is available at no direct charge to USDA personnel. Ongoing CALS search profiles are maintained that cover the AGRICOLA file of the NAL and such files as *Biological Abstracts, Bio-Research Abstracts*, and the file of the Commonwealth Agricultural Bureaux, which is composed of multiple abstract journals.

The PVPO's approximate location to the campus of the USDA's Beltsville Agricultural Research Center facilitates consultation with plant scientists who are leaders in their fields. This is particularly helpful now that budget restraints have curtailed the publication of many of the technical bulletins and publications, which the PVPO so heavily depended on in the past.

Minimum distance decisions have been made by the PVPO and examiners from the time the first applications were examined for novelty. Examiners consider the reliability of the data compiled from publications as well as the data presented by the applicant in determining the appropriate minimum distance for a character in assessing novelty of the cultivar. The PVPO has studied the reliability of characters and the appropriate distances to use for differentiating cultivars, with the advice of the Plant Variety Protection Advisory Board, the USDA-ARS, professional societies, and individual scientists. This assistance is continuing with the American Seed Trade Association appointing committees to study minimum distance in several crops. We expect their committee reports to be useful in further defining appropriate minimum distances.

In addition to verifying novelty, examiners must verify that a cultivar is new, uniform, and stable, and not a bacterium, fungus, or first generation hybrid. They must also verify that the application cultivar is eligible and is the property of an eligible applicant.

The PVPO has issued 2153 certificates of protection, with 11 certificates abandoned or withdrawn at the request of the owner. No certificates have been successfully challenged and defeated. The PVPO presently has 274 pending applications, which results in an average of 1¼ yr to issuance of certificates. Some of the delay is attributable to the applicants' delay in responding to requests for additional supporting information and extensions granted to applicants to provide the requested supporting information. Applicants have 6 months to respond to requests for additional information and may request time extensions of a month to several years to develop the needed support for their application. They also have 1 month to pay the certificate fee and return the form requesting a certificate.

PROBLEM AREAS

The crop exemption, or farmer's exemption, granted in the PVPA generates more complaints to the PVPO, and more requests for information and assistance than any other portion of the law. The Plant Variety Protection Board has studied the farmer's exemption and recommended that the amount of seed exempted be further defined. The Department of Agriculture has studied their recommendations and concluded that the exemption cannot be further defined by regulation because the PVPA's authority to write regulations is limited to the conduct of proceedings in the PVPO. The rights granted under the PVPA and the farmer's exemption are not enforced by the PVPO, therefore the USDA does not have authority to further define the exemption by regulation. Although the PVPO does not have enforcement authority, we do spend time answering questions related to the crop exemption. Some question the fairness of the PVPA crop exemption in contrast to the lack of a crop exemption in the Patent Act. This is especially difficult to answer from our interpretation of PTO policy, that all of the cultivars having received PVPA certificates of protection would have been eligible for utility patents under present patent policy.

Another source of objections and questions is the costs covered by fees charged to applicants. The PVPO fees are set to cover the full cost of the program, including administrative overhead. The objectors state that the PVPA provides some public good, and therefore should be partially funded through appropriated funds. In addition, they contrast the full fee funding of PVPA with the Patent Office plan to become fully fee funded in 1996; however, appropriated funds will continue to fund patent administrative overhead.

SECTION II

An Evaluation of the Issues, Challenges, and Opportunities Related to Plant Patenting

Part B. Plant Patenting Workshop Report

1 An Evaluation of the Issues, Challenges, and Opportunities Related to Plant Patenting

Executive Summary

The workshop on *An Evaluation of the Issues, Challenges, and Opportunities Related to Plant Patenting* met in Anaheim, CA between 31 Jan. and 3 Feb. 1989. The invited attendance included 78 individuals of broad disciplinary and professional interests. Following initial presentations that described biological, legal, and administrative aspects of intellectual property rights (IPR), the workshop broke into five working groups for detailed discussions on IPR impacts on research and development, socio-economic factors, and international genetic resources exchange and utilization.

The charge to the workshop was to "elucidate the key issues" of intellectual property protection of plants. The USA has long issued "plant patents" on asexually propagated varieties under the Plant Patent Act of 1930 and "plant variety certificates" on sexually propagated varieties under the Plant Variety Protection Act (PVPA) of 1970. Since *In re Hibberd*, a 1985 case, it also has issued utility patents on plants.

The workshop considered several questions and concerns that have been raised about utility patents; it also dealt with specific questions about the scope and characteristics of intellectual property coverage. The emphasis of discussion was placed on utility patents, since it is this coverage, and its relation to the PVPA system, that pose the most important and novel issues. The workshop did not take formal positions and nearly every point was disputed by some participants. Nevertheless, there were surprisingly high levels of agreement on many points. This summary attempts to record those levels of agreement, as well as review the important issues.

Among the broad issues and concerns, the most important is whether utility patents on plants are likely to be useful in encouraging innovation that would benefit the U.S. agricultural community. A large number of workshop members agreed that, save for a few questions and concerns, such patents would be beneficial, and that the value to the U.S. biotechnology industry would be shared by the farmer and the consumer. It noted, for example, that the number of private sector soybean breeding programs has significantly increased since the 1970 PVPA, and anticipated a similar pattern under the emerging utility patent regime. Thus, although there was some disagreement, the workshop was quite supportive of plant utility patenting.

Moreover, few in the workshop anticipated special problems for the small farmer or the family farmer. Most participants found no basis for estimating that the technologies protected by patents would be any more or less favorable to any particular scale of farming than those not protected by patents. The availability of patent protection may permit innovators to avoid the incentives toward vertical integration created by a desire to protect their technology through trade secrecy in the absence of patent protection.

The group was divided in their opinions about whether plant utility patents would increase concentration in the seed industry. It also was not obvious that concentration in the seed industry, as has been in process for a long time, is driven by potential or realized benefits of plant utility patents or PVPA, rather than other factors. They noted that, for a patented seed to succeed in the marketplace, any increase in prices associated with proprietary protection would have to be matched by a more than corresponding benefit to the farmer. Nearly all participants strongly urged that public sector breeding and basic research programs be maintained; this, along with antitrust law, would protect the farmer from any failure of competition.

On net, the group did not believe that there would be unfortunate environmental effects from the expansion of plant intellectual property protection. Improved intellectual property protection is likely to encourage development of genetically engineered plants that permit significantly decreased use of chemicals in agriculture or permit the substitution of environmentally benign chemicals for those associated with environmental problems. An area of possible concern, however, is whether there may be an increase in genetic vulnerability.

The workshop participants expressed uncertainty as to the impact of plant patenting on the domestic and international flow of germplasm. Access to state-of-the-art germplasm is essential both to research and to the public sector's ability to train future scientists. It was noted that many U.S. researchers—in both the public and the private sectors—are now exchanging material less freely and with more formalities than they once did. It was not at all clear to the participants, however, whether this trend was the result of utility patents, or of PVPA. Rather, it appears to be the result of commercialization of the entire sector. The participants emphasized the fact that patents permit germplasm exchange that might not occur under a trade secrecy regime.

On the international dimension, the trend toward patenting is likely to increase the value and, therefore, the preservation of germplasm. Nevertheless, the workshop recognized the risk that some developing nations will respond to the growth of intellectual property protection in the developed nations by decreasing their willingness to permit the export of genetic materials. The relative importance of patent issues in such a developing nation's decision is hard to evaluate in the face of preexisting export restrictions and a growing interest of developing nations in intellectual property legislation. The workshop strongly recommended continued and increased support for germplasm conservation and characterization efforts. It also recognized the need for increased efforts toward uniform international IPR in both developed and developing nations.

The workshop also dealt with a number of specific technical patent issues. First of all, nearly unanimously, it urged the narrowing of the "farmers' exemption" (PVPA § 113, 7 U.S.C. § 2543) under PVPA so that farmers would no longer be permitted to sell protected material for seed use. They would, however, continue to be permitted to use seed of PVPA varieties that they produced on their own farms. This modification to the PVPA law, which is generally rather favorable to the farmer and the small breeder, would make this law a more attractive option to some seed companies and would be consistent with developments in relation to international IPR.

Second, the workshop called for clarification of the "experimental use" doctrine under the utility patent laws. Part of the benefit of the utility patent disclosure is in providing a rapid advance of the state-of-the-art; this is encouraged by permitting liberal opportunity to reverse engineer or to design around patents, so long as the original patent holder maintains control over any components of its invention that may be embodied in a modified or improved product. Some believe that the exact status of the law in this area is not clear and fear that this uncertainty may have a dampening effect on research.

Third, a considerable number of the workshop participants were concerned by the question of patents on traits. The issue is exemplified by the situation in which an inventor finds a way to improve a plant and claims not only all plants within the same species (or related species in which the same improvement is possible through application of the disclosed method), but also claims all plants within the relevant species that contain the same improved characteristic, regardless of how the improved characteristic was generated. Thus, a patent might claim all plants of a particular species that are resistant to a specific pesticide, regardless of whether that resistance was obtained by the disclosed method or by a quite different method. Although some members of the workshop believed that overly broad claims were being granted, most believed that the statutory standards of utility, novelty, and nonobviousness would prove to be applied in such a way that claims of broad magnitude would be granted only when appropriate to the contribution made by the inventor. It was noted that some of these judgments may be made in the courts as case law.

Most participants in the workshop were unconcerned by the perceived possibility of patents on unmodified genes. In large part, these workshop members concluded that this issue derives from a misperception. Under current patent law, genes, as found in nature, cannot be patented; they can be patented only as isolated, modified, or inserted into a new host, etc. Hence, the possibility of patents on genes, is not nearly as broad as frequently perceived. Although there was some sense of concern at the possibility that the source of raw material for such a modified patentable gene might be an international gene bank, the majority of the workshop participants believed that the existing patent rules reach just about the right distance in permitting the patenting of genes.

There was one major issue on which the workshop was nowhere near consensus—that of whether or not a utility patent does or should reach progeny saved for planting, or, put differently, whether there should be a farmers' exemption for such patents, as there is for PVPA. It was recognized, however, that the breeder would have little incentive to sue his/her customer over limited use of the material on the farm. Thus, even if patents are given a broad scope, the practical issues are very much like those under plant variety protection.

Finally and most importantly, the workshop strongly urged additional discussion of and education about patent issues. Many of the fears of the plant breeding community are based on inaccurate perceptions of the utility patent system. The areas in which there remain concerns are areas that are, in large part, still subject to development and interpretation in the directions desired by breeders.

2 Workshop Rationale

INTRODUCTION

A workshop entitled *An Evaluation of the Issues, Challenges, and Opportunities Related to Plant Patenting*, was held in the Anaheim Hilton and Towers Hotel, Anaheim, CA, from 31 Jan.–3 Feb. 1989. The detailed program appears in Appendix 1. Assistant Secretary for Science and Education, U.S. Department of Agriculture, Orville G. Bentley, opened the workshop with a background statement and a challenge to the participants to clear thinking, open discussion, and fair treatment to all of the issues. His remarks are part of this report.

There are uncertainties about the practice of the law related to intellectual property rights (IPR) and the realized potential impacts of the relatively new opportunities in patenting plants through plant utility patents. Because of the great importance of proprietary protection of intellectual property in American agriculture and in international trade, Assistant Secretary Bentley very appropriately commissioned this evaluation.

The Crop Science Society of America (CSSA), along with its sister societies, the American Society of Agronomy (ASA), and the Soil Science Society of America (SSSA) agreed to coordinate the evaluation. The American Society for Horticultural Science and American Agricultural Economics Association joined in cosponsorship. The evaluation was designed by a Steering Committee including representation from the professional societies mentioned above, the USDA, industrial biotechnology, and the legal profession.

This report is based upon enthusiastic participation of nearly 80 individuals, but is especially due to work group chairs, facilitators, and rapporteurs who diligently prepared synthesis and summary from diverse views on a wide range of topics. Special credit is due to John H. Barton and Laura Meagher for their expert editing. This report is the product of the entire membership of the workshop, and, while there were divergent opinions among the participants, this report attempts to capture the sense of discussions and illuminate topics requiring further in-depth study. The individuals may not endorse all of the statements made but were given an opportunity to review and respond to drafts of the report. We believe the report represents a statement that all workshop participants "can live with."

Financial support was provided by the USDA, ASA, CSSA, and SSSA, and the institutions sponsoring participants at the workshop. The Steering Committee compliments David Kral and Cathy Goudreau at the CSSA Headquarters Offices for providing their exceptional organizational skills throughout this study.

GOALS AND OBJECTIVES

The overall goal of the evaluation was "to provide background on the concept of intellectual property protection of living matter as it relates to plant germplasm and plant products." Specific objectives were:

- To elucidate the key issues of plant patenting related to the pros and cons of patents.
- To identify the interrelationships between biotechnology, patent law, and agricultural production.
- To clarify the differences among plant protection using plant utility patents, plant process patents, and plant variety protection.
- To identify opportunities for resolution of basic principles and policy issues relating to plant patenting.

METHODOLOGY

The Steering Committee was formulated and held a planning session in Washington, DC on 20–21 Sept. 1988. The approach taken to meet the objectives of the evaluation was to identify specific potential or realized impacts of IPR. These topics then formed the basis for selecting appropriate representation of disciplines and special interests by individuals who were knowledgeable and willing to participate in this evaluation.

First, the following questions were raised:

- For patenting plants, what provision, if any, should be made for off-spring research, and farmers' rights?
- For patenting material characterized by plant traits, what limits should be placed on traits or coverage?
- For patenting genes, under what conditions should they be allowed? What clarification is needed with respect to questions such as:
  - What genes are patentable—naturally occurring genes or only engineered genes?
  - Can genes transferred between distantly related species by molecular-aided techniques, be considered differently from gene transfers made by conventional sexual means?
- For patents on processes, should there by any limits on the types of processes, or product of processes, that can be patented?

Second, the IPR has potential impacts that were identified in three categories.

I. Impacts of plant patenting on research and development processes, including exchange of germplasm, property rights, international competitiveness, and public-private relationships.
   A. Biotechnology
   B. Plant breeding

II. Impacts of plant patenting on producers and the public good.
   A. Economic, social, and ethical issues

B. Plant genetic resources and environmental issues

III. Impacts of plant patenting on international trade, genetic resources exchange, and utilization by developed countries; role of international agricultural research centers, economic effects on developing country farmers and urban populations; also impacts of developing and developed countries' policies on germplasm exchange and utilization.

A thorough examination of these topics was done at the workshop by work groups assigned to investigate each of the five potential impact areas. To address these topics, specialists were selected representing diverse disciplines (agronomy, biology, ecology, economics, genetics, horticulture, law, sociology, philosophy, plant breeding, and plant physiology); special interests groups (environmental, farmer, and consumer); private industry (biotechnology, seed companies, law firms, and seed certification); international agricultural development and trade; and last, but not least, the U.S. Patent and Trademark Office and the Plant Variety Protection Office. The workshop included 78 invited participants (Appendix 2) who were assigned to work groups such that each topic had a very broad representation of the above-mentioned fields (Appendix 3).

Each work group examined the questions regarding appropriateness for plant utility patents on varieties, genes, traits, or processes and then discussed the topic assigned to their work group. The work group chairs, facilitators, and rapporteurs were responsible for summarizing the discussions and preparing reports that were used to develop this summary document. Individual work group reports are available from the CSSA Headquarters Office in Madison, WI. Each group proceeded with their assignments independently, with the general instructions of developing a clear, concise statement of the issue; listing the important causes or factors contributing to the issue; describing the most important steps that can be taken to resolve the issue; and what agency, organization, or individual should take the lead in carrying out each step.

To set the stage for discussions, five papers were invited for both oral and written presentation (Appendix 1). These papers gave factual aspects of the biological basis for plant breeding and legal aspects of IPR.

The workshop could not resolve the many issues, but served to evaluate them and highlight need for further discussions and actions. It should be noted that the discussion was primarily directed to the impact of these patent issues upon the USA. The results of this workshop have been primarily of an educational nature and have established a basis for continued dialogue and future action.

Because the participants were a selected group of specialists and there was a need for broader discussion, a public forum and symposium on 4 May 1989 in Washington, DC presented the results of the workshop and several general information papers. No doubt this will be followed by further meetings and discussions throughout the USA.

3 Implications of Plant Patenting for Science, Technology, and Agriculture

Orville G. Bentley

Assistant Secretary for Science and Education
U.S. Department of Agriculture
Washington, DC

The greatness of U.S. agriculture—our abundance of food, fiber, and other agricultural products—stems not only from fertile soils and favorable climate, but also from the wealth of plant genetic resources from which scientists have selected or developed our successful crop varieties. Yet the fact that about 99% of U.S. cropland is seeded with varieties having their origin in other parts of the world is often not widely realized or appreciated. The germplasm from which these varieties of vegetable, fruit, nut, field, and industrial crops were derived has been accumulated over time through explorations and exchanges.

We are preserving these genetic resources in the working and base collections of our National Plant Germplasm System. The value of these collections to current and future productivity through improved varieties is based on their ongoing availability and accessibility to the plant breeders and research scientists of all nations. In recognition of the importance of genetic resources from all parts of the world, the USA, along with many other nations, supports a basic policy of free exchange of plant germplasm.

In recent years, however, there has been a growing trend toward protectionism in plants. In 1930, the Plant Patent Act (PPA) was passed to permit protection of most asexually produced plants. The PPA provided shelter for a single genotype but did not discourage the development of similar material with other novel traits.

Forty years later, in 1970, the Plant Variety Protection Act (PVPA) provided protection for seed-produced varieties. It sheltered the common genotype or variety, but this did not exclude the right of others to conduct research on it to develop new varieties, nor did it prohibit a farmer from saving seed of the protected variety for himself/herself or for sale to his/her neighbors.

In September 1985, in *Ex Parte Hibberd*, a Patent Office examiner rejected the application for a utility patent on a plant under the grounds that plants were traditionally not protectable under Section 101 of the Patent Act (a position that had been held since 1836), and that protecting plants under general U.S. patent laws would be a violation of UPOV (International Union for the Protection of New Varieties of Plants, 2 Dec. 1961, Revised Geneva 1978) of which the USA is a signatory.

The Patent Office Board of Appeals and Review subsequently reversed that Patent Office decision. In the reversal, the board rejected the argument on precedent by citing the 1980 Supreme Court decision (*Diamond v. Chakrabarty*) in which a patent was granted on a genetically engineered microorganism. The board also pointed out that a Congressional Act takes precedence over an international agreement such as UPOV (which has never been ratified by the Senate). This reversal opened the door for future utility patents on plants and brought with it a host of possible implications for research and agriculture.

Unlike patents under PPA and PVPA, utility patents may protect a single trait in a particular plant, and thereby offer much broader protection than PPA or PVPA. For example, the owner of a utility patent on a single trait in a soybean could exclude others from commercializing any soybean variety that possessed such a trait. On the other hand, utility patents are more difficult to obtain than PPA and PVPA. For instance, not only must the single trait in the soybean be novel, it must be unobvious to one of ordinary skill in the current state of the art.

Congress properly enacts patent laws to stimulate the commercialization of technology by protecting and rewarding the developer of the technology. The application of utility patents to improved conventional varieties may provide investment incentive and improved benefits for private companies, but does it assure reduced genetic vulnerability or improved production efficiency?

It seems obvious that private plant breeding companies must protect the research and development investment in their varieties if they are to remain in business. Protection and control on the sale of their varieties may be necessary to remain competitive. On the other hand, many plant breeders maintain that plant breeding products such as improved varieties and special genetic stocks do not need patent protection for immediate and widespread commercialization.

Over the past 20 to 30 yr, the number of private plant breeders has increased, while public plant breeding programs have remained static or decreased in size. The competition for resources and research opportunities in plant breeding has led some universities and state agricultural experiment stations to become more receptive to the income-generating potential of the protection offered by utility patents, thus raising new issues that may require reevaluation of guidelines and procedures.

We at the USDA are in the midst of just such an effort. Your workshop is part of that valuable reevaluation process. As we sort through the complex issues involved, it would be simplistic to expect a large and diverse organization such as the USDA to be monolithic in its outlook. Similarly, one cannot expect all universities to look at things from the same perspective.

There is real concern that focusing on plant patent protection and proprietary rights could interfere with the free flow of germplasm and genetic information, thereby killing the scientific goose that lays the golden egg. On the other hand, biotechnology-oriented scientists, whether in government, universities, or industry, tend to favor patents in plant breeding using biotechnologies that produce products not yet possible through conventional approaches.

We must also ask ourselves whether the patenting of breeding material in major crops may have a negative effect on the exchange and use of the protected material and on the research related to the traits in question. From a broad social, economic, and agricultural point of view, we need to assure that some small but important crops are not neglected. In specific cases, utility patents could provide the necessary economic incentive for research and development of a minor crop, although existing mechanisms, such as exclusive releases and licensing of public varieties, could also be effectively used.

Patent Office examiners could help meet these concerns about granting plant patents by demanding that applications rigidly meet the criteria of novelty and unobviousness. They should be permitted and encouraged to seek technical assistance from the plant breeding community on these criteria in order to demand clear evidence of inventiveness in the development of new products.

We must also consider that the widespread adoption of utility patents for new varieties may have foreign policy significance in terms of future U.S. access to world germplasm. It could bring about a reduction in the freedom of exchange of breeding materials and genetic information among plant breeders around the world, and be potentially counterproductive to U.S. and international agriculture. Already, many nations are becoming increasingly protective of their germplasm resources and are restricting exploration and access.

You are gathered to review and evaluate the issues, challenges, and opportunities related to plant patenting. The thrust of this plant patenting workshop is to look at the varied implications of applying utility patents to plants. The results of this workshop can help establish the policy, action, procedures, and/or legislation needed to assure that patents are used appropriately to provide needed incentive for the development and utilization of new technology, and are limited where they may provide a disincentive to progress. In your deliberations, you need to remain aware of the long-term economic, agricultural, and international significance of any proposed actions.

You have some hard work ahead of you for the next few days. I urge you to roll up your sleeves and tackle the tough questions, always remembering that we are here in good faith to not only air our differences, but attempt to reconcile them. The bottom line is, we need a policy that will do the most to keep American agriculture competitive in a rapidly changing global economy.

4 Workshop Summary Report

RELEVANT FORMS OF LEGAL PROTECTION

Although it would be inappropriate to present a lengthy dissertation on the different forms of protection available for intellectual property associated with plants, it is useful to provide a brief review. There are three relevant federal statutory systems; state trade secret law may also be relevant.

The statute under which coverage first became available is the Plant Patent Act (PPA), enacted in 1930 and codified at 35 U.S.C. §§ 161 ff. With certain exceptions, this law authorizes a "plant patent" for only asexually propagated varieties. (The term variety is considered to be synonymous with cultivar.) The new variety must be distinct. This statute is administered by the U.S. Patent and Trademark Office (PTO). (Although the term "plant patent" is sometimes restricted to this specific form of legal protection, it will often, for the sake of convenience, be used in this report to cover any form of intellectual property protection available for plants.)

The second such statute is the Plant Variety Protection Act (PVPA) of 1970, codified at 7 U.S.C. §§ 2421 ff, under which a plant variety protection certificate can be issued for sexually reproduced varieties. The new variety must be distinct, uniform, and stable. The breeder's right is limited by several provisions intended to protect farmers, to allow use of the protected material for breeding, and to allow farmers to reuse seed on their own land and, under some limited circumstances, to sell such seeds to their neighbors. This statute is administered by the Department of Agriculture's Plant Variety Protection Office. It also conforms with the International Convention for the Protection of New Varieties of Plants, administered by the International Union for the Protection of New Varieties of Plants (UPOV), of which the USA is a member.

The final statute is the oldest of all, the basic Patent Statute, codified at 35 U.S.C. §§ 100 ff. The PTO has granted utility patents for processes for producing plants for some time; it moved further in response to the 1980 U.S. Supreme Court decision in *Diamond v. Chakrabarty*, 447 U.S. 303, which held a bacterium, as a living organism, to be patentable. Beginning with *Ex Parte Hibberd*, 227 U.S.P.Q. 443, in 1985, the PTO has been issuing utility patents for plants, provided they are useful, novel, and nonobvious. This statute does not have the exemptions for farmers or for other breeders as included in the PVPA, although it does allow for a judicially created experimental use exemption whose limits are not completely clear. This statute also authorizes the patenting of individual genes, genetic processes, and plant characteristics under some circumstances. The scope of such patent protection of genes, processes, and specific plant characteristics is probably as important an issue as the actual availability of patent protection on plants.

It is the patenting of plants under 35 U.S.C. §§ 100 ff that has created new opportunities for protection of intellectual property relating to plants. This was given major attention in this evaluation. It is important to distinguish patents on processes or methods of plant germplasm manipulation (plant process patents) from patents on genes or traits (plant utility patents).

Finally, there are forms of protection available even when these statutes do not apply. The most obvious practical example is the use of hybrids in which physical and contractual control of the parental inbred lines amounts to the equivalent of trade secrecy protection on the hybrid product. The law supports this control process through rules for the protection of trade secrets. These rules do not prohibit reverse engineering or parallel inventions, but do prohibit, for example, direct theft of the protected material or its misappropriation by employees.

FUNDAMENTAL ISSUES

Following the subjects addressed by each of the work groups, there were four areas of fundamental importance that are influenced by intellectual property protection of plants. The substantive points of discussion are summarized here.

Effects on Innovation and the Agricultural Supply Industry

One of the key issues put before the workshop was whether innovation in plant breeding research is effectively promoted through intellectual property rights (IPR). Several work groups considered this issue. In particular, they explored the implications of IPR for division of research between public and private sectors and for information flow. Patents and their effect on genetic resources availability might, for example, have impacts on structural aspects of agriculture, such as university-industry relationships and the number and size of seed companies.

Level and Orientation of Research

The economic analysis of IPR recognizes first that in the absence of IPR, incentives to research and development (R & D) in private firms generally depend on the ability of a firm to maintain inventions in secrecy and to achieve lead time over competitors. Introducing IPR into this situation is expected to create incentives and disincentives.

Incentives:

Incentives to undertake R & D are expanded. This expansion takes three forms. Existing firms undertaking R & D may increase R & D. Firms not currently engaged in R & D may initiate R & D activities. New start-up firms, based on R & D programs, may be instituted.

Inventions are disclosed and thus removed from secrecy. This promotes further discoveries because disclosure reveals concepts and techniques useful to further invention.

Disincentives:

Inventions may not be used to their optimal efficiency because IPR allow the holder to limit their use to a degree greater than the marginal cost of replication of the invention. This loss applies only to inventions that would actually be made in the absence of IPR; a parallel loss arises for inventions protected by trade secrecy.

The scope and strength of IPR will affect the balance or potential tradeoff between these incentives and disincentives. Strong protection increases incentives for innovation, and can also increase disincentives. Present utility patent systems are generally regarded as achieving a reasonable balance in most fields of inventions and thus stimulating useful innovation.

Experience under the 1970 PVPA appears to confirm that IPR can stimulate significant private sector plant breeding activities. For example, the number of private sector soybean breeding programs has increased from less than 5 prior to the PVPA to more than 30 after PVPA implementation. Annual PVPA certificates granted to private firms have grown substantially over the past dozen years. About 7% of these certificates have been granted to foreign firms and 10% to public research institutions (USDA-ARS and state agricultural experiment stations). Many varieties are on recommended variety lists and are planted widely.

There is not yet enough experience to establish how much plant breeding activity has expanded because of utility patent protection, since such protection has been available for only a limited time. Biotechnology R & D has shown high rates of growth and the firms involved generally regard the potential values of IPR as important to this growth. However, patents are generally expected to be good for companies and economic competitiveness, and to increase the amount of funds available for their research. Moreover, the availability of prior art findings should enhance additional plant improvement and create further incentives to identify and develop unique germplasm.

Consideration was given to the possibility of adverse impacts on small firms that have traditional applied breeding programs, and potentially face additional costs and legal concerns because of patents. Nevertheless, the fruit and seeds of their efforts will be better protected. Also, new entrepreneurial opportunities will be created for small firms that specialize in narrow fields of genetic research or participate in combining, trading, or brokering the wider range of genetic materials that will be available to meet breeder's objectives.

Public-Private Sector Relationships

Overall, the group agreed that plant patents would be likely to expand the availability of funds for plant research in both companies and universities; this was acknowledged as a positive factor. However, even as part of an overall trend of R & D growth occasioned by the commercial promise of biotechnology, the pursuit of external funding by universities, and national emphasis on economic competitiveness, plant utility patents can contribute to changes in the relative orientations of public and private sector research.

Although small public breeding programs will probably be affected similarly to small private breeding programs, overall impacts on the public sector research system are less clear. Very generally summarized, concerns were expressed that basic research might be undervalued and underfunded compared to research that is more immediately applicable in terms of financial return. Moreover, utility patents may change the nature of public agricultural researchers' goals, increasing the importance of short-run as compared to long-term goals that are most likely to lead to technology that can contribute to international development.

Several factors were identified that might intensify these impacts. Chief among these were that universities may be motivated to obtain funds from patent royalties hoping to offset declining state and federal funds.

At the same time, on the positive side, it was noted that royalty earnings can be allocated by universities and other public research institutions to long-term goals consistent with their missions (more than is possible with external grant funds). Furthermore, firms do at times invest in basic research, as long as they are assured of licensing access to any patentable discoveries that should arise.

Public breeders also expressed concern about their ability to retain publicly developed varieties in the public domain and expressed a fear that they would have to resort to defensive patenting to keep the new materials from being patented by the private sector. It was recognized that a patent is unavailable in the face of prior publication by other researchers or for a product copied from other researchers. It was also recognized, however, that a privately patented development could foreclose a public sector development if the two were essentially similar and developed independently at approximately the same time.

The workshop strongly urged continuing support of public sector research. It emphasized that education of future scientists has been a major mission linked with public institution research programs. If there are not active breeding programs at universities and within the USDA, will there be sufficient numbers of trained practitioners in the future?

Information Flow and Germplasm Exchange

Plant utility patents affect information flow positively in the long-term, both because of the requirement for disclosure to the point of enablement, often including deposits of the living plant and seeds, and because inventions enter the public domain after expiration of the patent. However, publication can be slowed through the need to avoid public disclosure prior to filing for the patent. Even though there is a 1-yr grace period in the USA that permits a patent application to be filed within a year after publication, inventors usually delay publication until after the application, in order to protect their rights under foreign patent systems. In contrast, trade secrets (which are sometimes the likely alternative) always inhibit information flow. Thus, plant utility patents may have a short-term negative effect on sharing of germplasm and information, but long-term benefits are expected.

The effect of utility patents on the exchange of germplasm within the USA received particular attention and concern. Private firm to private firm exchange may be increased because the patent owners can protect investments and the opportunities for cross-licensing are positive. Public-to-private exchanges are likely to increase, however, they may become much more formalized.

Public-to-public exchange will likely also become more formal. There was special concern here, however, particularly with the possibility of a changed "mind-set" of public-sector researchers toward increased secrecy. Public institutions are concerned about access to germplasm in general, and have particular concerns about access to offspring of patented varieties. Clearly, the exchange of more advanced breeding lines has decreased and this is seen by some as a result of the changes brought about by plant variety protection. Materials that are exchanged are likely to be less developed and lack any evident commercial value. Exchanges, regardless of the parties involved, will increasingly include memoranda of understanding and document future responsibilities of both parties. It was also suggested that the length of time to complete paperwork would involve delays in exchange and access.

The perceptions that plant utility patents will diminish germplasm exchange cannot, to date, be verified because of the limited extent of patent activity so far. Initial patent applications in the public arena have generated grave concerns among some breeders about their possible effects on reciprocal exchanges with the patent applicants. There was general agreement, however, that patents are one of several forces (although by no means the only one, nor in some cases, even the most important force) currently increasing the value of plant innovations and/or the ability to appropriate that increased value.

Possible resolutions of the problems of restrictions on exchange of genetic material include the following. First, there is a need for education, particularly by organizations that have a strong interest in maintaining the exchange of genetic resources. Second, some suggested that an amendment to the patent law to clarify the research exemption (discussed later in this report) would contribute to the exchange of genetic material by increasing the flow of patented material into public and private plant breeding research programs.

Another possible set of resolutions would involve increasing the moral or financial incentives to encourage free exchange. Thus, land grant universities might have a common policy of exchange of genetic material, which endorses free exchange. It was also suggested by some work groups that it may be useful for federal law to stipulate that universities, public or private, receiving Cooperative State Research Service (CSRS) competitive grants or other USDA funds, should adhere to free exchange.

Several other recommendations were made in response to the information flow and germplasm issues: (i) educate potential inventors as to prompt filing and the possibility of publication thereafter (or even as much as a year before filing if only U.S. patent protection is of interest), (ii) continue germplasm maintenance activities in universities and the National Plant Germplasm System, (iii) examine when and how publicly owned germplasm will be available to agribusiness under the different mechanisms of intellectual proprietary rights, (iv) educate the germplasm community about the realities of utility patents.

Effects on Farming and Consumer Communities

Direct Impacts on Farming and Consumer Communities

The IPR-induced inventions will have complex long-run impacts on farm incomes and farm adjustments. They may place further adjustment stress on most farm sectors because of their supply-increasing impact. Moreover, IPR-induced inventions will probably be management responsive and increase the premium earned by good farm managers. Whether or not this provides an incentive for increased farm size, it certainly provides an incentive for specialization, and thus furthers recent trends in U.S. agriculture. While U.S. agriculture should be able to absorb these possible economic changes, it may be much more difficult for a farmer to survive major fluctuations in the agricultural economic system. Moreover, IPR-induced inventions will probably not be developed in an even fashion for all regions, and their regional effect cannot be easily predicted.

Most important, however, IPR-induced inventions are expected to provide farmers with a broader range of input options, as in insect control, and output options, as in crop products. These inventions are generally priced at less than their real benefit to the farmer, even when their prices include returns to private-sector R & D advertising, licensing, and other costs. Farmers will purchase technology if it enhances profits or ensures marketability of products. Thus, the net effects of IPR on agriculture as a whole can be expected to be positive.

Effects on Crop Consumers

Consumers will be expected to gain because of a broader range of plant quality options and lower prices, where patented material is superior and results in lower costs of production. These gains will be favorably distributed throughout the economy. They are available to everyone and will be particularly beneficial to low-income families who spend a higher percentage of their budget on farm products than higher income families.

Concentration in the Supplier Sector

There has been concentration in the seed industry due to mergers. Moreover, IPR may threaten the existence of some small firms that do not have access to large research programs. It is not clear, however, that the mergers are primarily attributable to IPR, and indeed there is evidence that many new companies have been started in response to IPR.

Ethical Issues

Some workshop attendees perceived ethical problems with patenting life forms. These include concerns about patenting the food crops that sustain us and about limiting the distribution of germplasm. The workshop was unable to address these issues as philosophical concerns; it did, however, acknowledge these concerns as deserving further discussion and resolution.

Environmental and Domestic Germplasm Issues

The workshop recognized that the integrated balance of managed and natural ecosystems is at the heart of human survival and that an ultimate test of all human undertakings should be the assessment of long-term overall ecological balance. The biological world is in constant flux and balance is not a fixed point but a moving equilibrium. Biological stability is maintained by a vast diversity of organisms using a cyclic flow of resources. Some of the most successfully managed ecosystems have elements that mimic the diversity-stability equilibrium found in natural systems. A good measure of how well the system works is not when everything is right, but rather how large the disruption when things go wrong.

There is uncertainty about how proprietary rights will affect germplasm availability, as discussed earlier.

Patenting may speed development of methods of biological control, which could reduce the magnitude of environmental problems caused by current pest control methods. This will be an important factor for moving towards greater sustainability. No explicit conclusions emerged on genetic vulnerability. It was noted, however, that genetic diversity may be increased, thus reducing genetic vulnerability because of the increased incentives for plant breeders to invest in acquisition and development of new proprietary material.

Since the best environmental safeguard for agriculture worldwide is the universal availability of diverse germplasm, the workshop urged that the following actions be taken.

- Conduct research on the impacts of patenting on availability and development of genetic resources for plant improvement.
- Provide educational programs about the necessity for accessibility to germplasm.
- Evaluate and develop, if necessary, policy which ensures that accessibility.
- Strengthen PVPA to enhance its utilization as an alternative to plant utility patents.
- Improve characterization and evaluation of material in genetic resource banks.
- Ensure that effective IPR are available to encourage private investment in research.

International Issues

Competition and Intellectual Property Cooperation among Developed Nations

The IPR-induced inventions are likely to benefit the competitiveness of both the U.S. seed industry and production agriculture by providing a comparative advantage for the USA in the international marketplace, and generating opportunities to expand exports to traditional and possibly new markets. The development of proprietary varieties in different nations has already led to increased trade in such varieties. Thus, the list of plant variety protection certificates issued from 1971 to 1986 shows that certificates were issued to foreign organizations for barley, several grasses, and some vegetables. Plant utility patents will probably provide further opportunities for both transnationals and foreign corporations to further participate in the global market. Similarly, varieties developed under the PVPA have been exported and the adoption of utility patents will probably increase the products available for expanding this potentially important market.

It should be noted, however, that the interests of the U.S. breeding material export industry are not always the same as those of U.S. production agriculture. Increased export of seed to specific markets, for example, of soybean seed to Brazil, will, in turn, affect export opportunities for U.S. products such as soybean. Nevertheless, the consumer can benefit when advances in plant breeding support more efficient foreign production; moreover, it is clearly essential to maintain the competitiveness of the U.S. plant breeding industry. Foreign farmers will buy improved varieties whether or not they are sold by U.S. firms.

There may also be an impact of utility patents on international exchange of germplasm, depending on the nature of property laws in exchanging countries. Inaccurate perceptions of patent rights, particularly the failure to recognize that publication generally precludes foreign patenting and that a patent prohibits unauthorized use only within the nation in which it is issued, may reduce exchange of "unfinished" germplasm in the short-term. However, variety exchange should be enhanced as foreign scientists become aware that the deposit of seeds under plant utility patents means that seeds should be available for research use. As noted later, the matter of research exemption under plant utility patents needs clarification.

To ensure an effective international market in breeding materials and an effective international exchange of genetic materials, efforts may have to be undertaken to harmonize criteria for granting IPR and the scope of IPR, particularly with the European community UPOV members. Proposals for the revision of the UPOV Convention are currently underway internationally and could lead to a Diplomatic Conference to revise the convention in the early 1990s. The absence of patent protection in foreign countries discourages exports of U.S.-patented products to such countries, as well as R & D and manufacturing of the U.S.-patented products in those foreign countries that do not provide protection. In a global economy, in order to increase R & D, it is important to encourage worldwide protection of IPR.

The workshop thus suggests that the USA should encourage the same level of protection overseas as is available in the USA. Leading this effort should be the appropriate agencies within the Department of Agriculture, Commerce, and State, and the President's Science Advisor. Universities and the private sector should be encouraged to build support among their parallel constituencies within foreign countries.

Developing Nations and International Germplasm Flow

The issues of germplasm access also extend to the developing nations. Germplasm has never been freely exchanged between all countries. Plant utility patents are generally perceived as having the potential, whether accurate or not, to increase the value of improved germplasm; this may increase developing nation reluctance to provide their genetic resources free of charge. Both public and private sector breeders are concerned that countries having valuable genetic resources will be reluctant to share them in the future.

Several factors may intensify the tendency toward a decline in the extent or speed of genetic resources exchange. Among the most important of these factors are geopolitical sensitivities, including the growing reluctance by many developing countries to allow genetic material to move across their borders and differences in national value systems with regard to giving proprietary protection to life forms. Another factor that may lead to a tendency toward a decline in the exchange of genetic material is that the International Agricultural Research Centers (IARC) do not distribute germplasm protected by proprietary rights without explicit permission from the owners of those rights.

Although the factors just described suggest that utility plant patenting may tend to reduce exchange of genetic material, there are several offsetting or compensating factors that should be considered. First, since plant patents do not exist in most countries, there is no legal basis for restrictions on the exchange of patented plant material in these countries. Second, there exists informal means to circumvent restrictions on the flow of germplasm (for example, informal relations among plant breeders). Third, the disclosure provisions of patents are tantamount to publication of the technical basis of patented innovations, which under some circumstances, can contribute to the flow of technical information to developing countries.

If developing countries perceive the economic aspect of patent protection as the major reason for developed world interest in germplasm, barriers to exchange can develop and germplasm can be lost by neglect. There is a serious risk that these leaders will define their policies based on misperceptions about patents. Intellectual property rights stimulate plant breeding, giving rise to more elite lines. Patenting may speed the discovery of "new" genes and gene combinations. Thus, it is important to provide information about the actual workings of the patent system.

It is also important—for the sake of both the developed and the developing nations—to maintain and preserve germplasm in gene banks, even though the patenting of plants may also decrease the flow of germplasm from developing countries to the international research centers. Material is already being lost in both developed and developing countries, with the largest fraction in developing countries. Improved germplasm ultimately displaces local material in agriculture so conservation of plant genetic resources is essential. Landraces will continue to be displaced and lost. The tragic aspect is that once lost, germplasm is essentially gone forever; it could be of benefit to all, whether from developed or developing countries.

Thus, proprietary protection of genetic material in the USA creates a responsibility for the USA to assist the developing countries and the IARC in building the capacity to conserve and make use of these genetic resources. The U.S. Agency for International Development (USAID) has been mandated by congress to focus upon this biodiversity issue while continuing its support of the IARC Consultative Group for International Agricultural Research (CGIAR). Through the CGIAR, the International Board on Plant Genetic Resources (IBPGR) receives support for its conservation efforts.

The recent Keystone International Dialogue on Plant Genetic Resources (*Diversity*, 1988, Issue no. 15, p. 6–7) clearly indicated that increased funding to support genetic resource conservation is needed. It further identified several mechanisms of funding, preferably through a United Nations fund, governed by a special board, to consider the most urgent needs. The dialogue further suggested various ways to assess all nations to contribute to this fund.

Thus, the workshop suggested the USA should play a strong role in supporting the concept of protecting global plant resources and arriving at an equitable assessment mechanism. It also recognized the strong impact that USAID-university partnerships have had for institution building in developing countries, especially through the Collaborative Research Support Program. These efforts might be increased to strengthen effectively these relationships with developing countries. Collaborative state and federal relationships with developing country programs might be enhanced in training scientists, technicians, and administrative personnel associated with plant improvement. This includes support for genetic enhancement. It also includes in situ conservation because wild relatives of crop plants present special problems that may be best served via in situ conservation. Moreover, in situ conservation complements gene banks by retaining biological information about interactions among species and their associated pests and beneficial species. In trying to get agricultural systems that closely mimic natural ecosystems, it is necessary to study the energy and material flows in remaining natural systems, again emphasizing the need for in situ conservation.

In addition, a suggestion from the workshop was that the U.S. government implement a national proclamation that germplasm be removed from the trade embargo category, thus reaffirming U.S. policy that germplasm is a freely exchangeable common heritage.

Use of Intellectual Property Rights by Developing Nations

One work group considered whether developing countries should implement intellectual property protection as a means of enhancing plant innovations that contribute to agricultural development. The work group noted that different nations have different approaches to these issues and asked if it is appropriate for small and/or poor developing countries to give much emphasis to implementing proprietary protection or "nationalization" of genetic resources.

Trade groups, the U.S. government, and international organizations may want to encourage the development of IPR for plant materials in those developing countries that have an adequate legal infrastructure and agricultural research systems to support such protection. It was felt by one work group, however, that for many developing countries, intellectual property protection of plant innovations would be inappropriate under their present circumstances. The work group recognized that this position is not entirely consistent with current U.S. trade law, which states that if trading partners do not have adequate protection of intellectual property, they may be subject to trade retaliation by the USA.

SPECIFIC ISSUES

A number of specific issues were considered by all the work groups; consensus emerged on certain ones.

Plant Variety Protection Act and Farmers' Crop Exemption

The crop exemption for farmers has been discussed broadly within the agricultural community for a number of years. Currently, the PVPA allows farmers such broad opportunities to sell protected varieties for use as propagating material that some participants felt that incentives for research on variety development are reduced; enforcement against farmers widely selling protected varieties has so far been only marginally effective.

The workshop was in consensus that the "crop exemption" provision of the PVPA (§ 113) should be modified to allow the sale for seed of a PVPA-protected variety by only the PVPA certificate holder or its licensee. This change would not be intended to restrict the farmer from saving seed of a protected variety under the PVPA for planting on his/her farm.

Other possible changes in the PVPA that were considered included, "tightening up" the research exemption as delineated by UPOV by changing rules for distinctiveness. Thus, when a variety that is essentially derived from another, the breeder of the derived variety might not be free to market the variety "without obligation to" the original breeder. Distinctness might also be defined in genetic terms by specific genetic markers, including molecular traits. In some instances the statistical measure, coefficient of parentage, may be useful. It was noted that changes of this nature are included in the UPOV revision proposal currently under discussion, which might, in due course, lead to a need to revise the PVPA.

Protection of Plant Varieties and Breeding Processes under Utility Patents

As mentioned previously, the workshop considered permitting plant utility patents for plants and noted the various criticisms that have been made. These include the concerns that there will be fewer small public and private institutions engaged in varietal development and that the nature of their activities will change, that the number of released varieties will diminish and that there will be a resultant limited choice of varieties for many small and varied agroecological niches in the USA, and that patenting varieties will reduce genetic advance and increase genetic vulnerability. In general, however, the workshop believed the long-term effects of utility patents on varieties would not affect plant breeding adversely and may well provide new opportunities not currently available. Similar reservations were voiced when the PVPA was implemented, but, in general, its effects on plant breeding have been positive.

Patents on breeding practices, that is, plant process patents, have been in use for a long time and were not seen as raising new issues related to the patenting of plants. Therefore, no recommendations for changes were offered.

Farmers' Rights under Utility Patents

It was unclear to the group whether farmers who purchase seed patented through plant utility patents (or seed containing patented genes or traits) can grow a second generation using the offspring seed. Farmers need to know that the limitations with plants protected by a plant utility patent may be different from the more familiar limitations under PVPA. Although farmers' rights to reproduce for their own use should be clarified, the actual allocation of specific rights between the patent holder and his/her customers can often be better addressed through license provisions than by legislative change.

Experimental Use Exemption under Utility Patents

The workshop agreed almost unanimously that clarification of the judicially devised experimental use doctrine is desirable. Although there was not unanimity about the precise direction of clarification, many urged that a plant protected by a utility patent be usable without restrictions by a third party plant breeder for purposes of conducting R & D activities leading to the development of noninfringing plants. This would not include the commercial sale of the patented plant or plants dominated by the patent, without agreement from the owner of the patent on the variety, traits, or genes used in the new development.

It was recognized that there are some conflicts between public- and private-sector institutions on the issue of germplasm and variety control. Private firms may prefer a utility patent with no research exemption for some applications, for example, inbred lines. If there is a research exemption, private firms might continue to protect their inbred lines with trade secrets.

The PVPA contains specific language on the research exemption that appears to satisfy the various concerns. Utility patent legislation, however, is silent on the subject. A limited, judicially based exemption exists, and indeed the right to experiment with patented products and processes is central to the spirit and philosophy of the patent system with its intent to promote science and technology. With living organisms, this opportunity to experiment may require access to and use of the invention itself. On this subject the case law is unclear and may currently permit only basic research as opposed to use expressly for variety development. For some participants, the uncertainty about the research exemption has a chilling effect on future R & D and appears to obscure the intent of the patent law.

There were some differences within the workshop over the way to resolve the issue; some saw no need to clarify the status of research use of patented plants or parts thereof, except as required by special situations through case law. Other members, however, suggested clarification, through statutory change to achieve the following:

> The public (including commercial breeders) should have the right to experiment with plants protected by IPR in such ways as could lead to the development of new and distinct products or processes. This right is independent of the intent of such experimentation, that is, whether the motivation is for knowledge or for profit. However, the sale of the newly developed product could infringe the initial patent(s) and may require licensing or other arrangements with the original patent holder(s).

Consideration was given to the alternative possibility of compulsory licensing of material protected under plant utility patents as a way to permit use of patented material for breeding purposes. In general, the workshop was not sympathetic to this approach. Compulsory licensing was viewed as a "taking" of the owner's property and thus an unjust interference in his/her property rights. While compulsory licensing would be beneficial to some plant breeders and firms in the short-term, the long-term effect would seriously erode research incentives. This erosion would reduce benefits of increased genetic progress that are otherwise attainable with utility patenting of plants and plant parts.

Patenting of Plant Traits

An area of concern to a number of participants was the patentability of traits. The particular concern was whether inventors, who have shown a particular way to obtain a plant with a new beneficial trait, should be able to claim protection for all plants (not necessarily limited to one species) possessing that trait, regardless of whether the trait was induced by the disclosed process or by a completely different process. Some suggested that, in the face of such patents, alternative genetic approaches to achieve crop performance progress may be inhibited or blocked; crop vulnerability was a particular example of concern. In contrast, others feared that limits on such claims could deprive inventors of the appropriate rewards intended by the patent system. A majority of the working groups supported the patenting of traits.

In response to the concerns, some suggested that the scope of the patent should be proportional to the disclosure and the state of the art. For example, it was suggested that claims for traits should be awarded only in cases in which the trait was not previously well characterized and/or precise isolation and transfer of the trait was not suggested as possible by those with skill in the art, and when the disclosure is sufficiently detailed to permit replication of the isolation and transfer without undue experimentation.

Protection of Natural Genes

Some initial concern was voiced regarding the patentability of genes. This is not an issue with respect to patenting of engineered genes. Such patenting was thought proper in view of the magnitude of the effort and costs involved in generating such inventions. Patent claims embodying naturally occurring genes, however, engendered initial concern over the possibility that it would place undue restrictions on that which was previously in the public hands.

It emerged that existing patent law provides limits that more than one of the work groups found appropriate. Claims must be equal in scope to the inventors' discoveries. For example, an enforceable patent will typically not be issued with claims directed to a gene per se that had already been identified (had been "known") to the extent that its manipulation, at a molecular level, would have been considered a reasonably predictable endeavor by those of ordinary skill in the relevant art. Moreover, a valid claim to a gene should not cover a previously available variety or other plant material which may contain that gene.

Conversely, a novel genetic construct may be patentable per se, as may plant material into which the construct is transferred to some useful end. In particular, a gene that is "discovered" (characterized molecularly for the first time) and made accessible to the art in a useful form would appear to be appropriate subject matter for patent protection.

Thus, the majority felt that as long as the law is applied to achieve fair and just results, no specific restrictions on the patenting of genes, either engineered or natural, are justified. It was felt that the additional economic incentive offered by patents, to both engineered and naturally occurring genes, is sufficient to justify the limited property rights provided by utility patents.

Foreign Intellectual Property Issues

In the USA there is a 12-month grace period that allows an inventor to file a patent application within 12 months following publication. This permits ideas to reach the public domain earlier, without barring patentability. Few foreign countries have such a grace period. It was suggested that the USA seek foreign adoption of a 12-month grace period for prefiling publication. The U.S. Trade Representative, the Department of Commerce with the Patent and Trademark Office, and the private sector could all assist in urging such change; the subject is presently under international discussion.

Encouragement of Support for Gene Banks

Genes conserved in banks are an important component of diversity, and hybridization with progeny selection and testing maintains the flow of genetic resources. Evaluation of genetic resources from gene banks and recycling of genes from elite materials are the fuel of the plant breeding process. The workshop believed that this germplasm flow can be strengthened to facilitate transfer and exchange.

As noted above, the concern that patents restrict germplasm availability, can be minimized by insuring that a strong gene bank system is available for collection, conservation, characterization, and evaluation. It should also be noted that a large array of germplasm is in the public domain today and patented materials will enter the public domain without any restriction on use when the patent expires. It is important to support better characterization of material in gene banks so that characterization is known and in the public domain, available, and therefore, not subject to IPR.

Special Issues for Public Sector Research

Public sector institutions, for example, the USDA-ARS, and state agricultural experiment stations, have the opportunity to protect intellectual property. For new plant varieties, plant genes, and breeding or biotechnology processes, a number of state agricultural experiment stations have chosen to use the various types of protection available; others and, for the most part, the USDA-ARS have chosen not to use IPR protection.

A public institution must consider a number of factors in making decisions on future directions in plant science genetics and breeding research, and on whether or not to seek protection for new inventions. If, for a particular crop, there is a strong private industry effort, one possibility is to restructure public institution programs to place more emphasis on complementary "pretechnology" science, such as basic research in genetics and extension-related activities, such as objective evaluations. Emphasis could also be directed toward crops for which there is little private effort. Another potential response is to continue with variety development, seek appropriate protection (certificate, patent, etc.), and license the technology. This is an appropriate direction for institutions with a commitment to education and training in gene manipulation and plant breeding.

Although no specific recommendation emerged as to action steps, several possible directions of resolution were identified. First, increased public funding of agricultural research would reduce pressure to emphasize short-term goals. Second, land grant universities should be encouraged to engage in strategic planning (to evaluate whether their goals are consistent with broader stated missions and to avoid short-term goals becoming predominant). Third, it is important that there be adequate funding for public plant breeding; public plant breeders should not depend on external funding.

Information Flow Needs

Throughout the session, it became clear that substantial confusion on IPR persists. The workshop saw two types of apprehension.

- Uncertainty that reflects an as-yet-unsettled legal situation.
- Misunderstanding of utility patent law and practice, as it exists, among parties influenced by that law.

The first kind of uncertainty is particularly relevant with respect to the likely scope of researchers' and farmers' rights. Clarification is necessary, but much of it will take place in the courts and the legislature as case decisions are made and legislation clarified.

There was also concern about the standards used in granting patents. The workshop noted the application of the nonobviousness requirements is changing with time, and that, as the PTO enlarges its experience and its data bases, concerns about patents issued that are too broad or too obvious will diminish. There was particular emphasis on the need for the PTO examiners to have continuing technical update training and have access to all pertinent technical teachings that are public knowledge. This concern arose, in part, because examiners cannot seek external technical assistance in making their determinations because of the confidentiality provisions of the law.

These issues are being addressed by the PTO. In addition to ongoing programs used for all technologies, the PTO is developing a program, in cooperation with outside organizations, to enhance its abilities of executing its mission for this new and rapidly growing technology area. The Patent and Trademark Office is forming a Biotechnology Institute that will provide enhanced training and access to technical information in cooperation with other government agencies, universities, and trade and bar associations.

With respect to the second type of uncertainty, there was a strong sense that many of the affected communities do not understand the actual patent system. There is a lack of knowledge on the part of inventors as to their obligations before they can get a utility patent, on the deciding criteria for granting utility patents, and on the possibility that the improper issuance of a utility patent can be corrected through a reexamination procedure. There is particular lack of comprehension about the standards for the patentability of naturally occurring organisms and the permissible scope of product/process claims. A better understanding is needed on the most appropriate means for the protection of intellectual property in the public sector. Although publication may be appropriate under certain circumstances, it is often not adequate to assure the protection of new technology that is deemed important to be made freely available as a public service. Publication may actually place ownership of a discovery in jeopardy, due to the 12-month grace period allowed in the USA prior to initiating a patent application. This is not true in foreign countries where publication constitutes an immediate barrier to patenting.

Education is the action step called for, at all levels. This may take several forms, perhaps including a small booklet, which could be produced by the PTO and the USDA, and the Societies represented in this workshop. At the least, a single-page clear statement issued by the PTO would be invaluable for many audiences.

A compilation of important audiences to be addressed and the agencies, organizations, or groups/individuals that might be involved in developing and implementing educational programs are presented in Appendix 4.

Appendix 1. Workshop Program: An Evaluation of the Issues, Challenges, and Opportunities Related to Plant Patenting

31 January–3 February 1989
Anaheim Hilton and Towers Hotel
Anaheim, California

Sponsors: American Society of Agronomy
Crop Science Society of America
Soil Science Society of America

Cosponsors: American Agricultural Economics Association
American Society for Horticultural Science

Tuesday 31 January

Evening Session

7:00 pm **Registration. (Capistrano A/B)**

7:45 **Welcome, introductions, and announcements.**
Calvin O. Qualset and Donald N. Duvick, Workshop Steering Committee Cochairs.

Keynote address.

Implications of plant patenting in the advancement of science, technology, and agriculture. Orville G. Bentley, Assistant Secretary for Science and Education, U.S. Department of Agriculture.

9:00 **Reception. (Palos Verdes A/B)**

Wednesday 1 February

Plenary Session (Laguna A)

John H. Barton, Stanford University, Presiding.

8:15 am **Introductory remarks. Goals for the workshop.** Robert F Barnes, Donald N. Duvick, and Calvin O. Qualset.

8:30 **Plant breeding, crop cultivars and the nature of genetic variability.** Frederick A. Bliss.

9:00 **Discussion.**

9:10 **Intellectual property protection and plants.** Sidney B. Williams, Jr. and Kenneth A. Weber.

9:40 **Discussion.**

9:50 **The patenting of genes that encode agriculturally important traits.** Stephen A. Bent.

10:30 **Break.**

10:50 **Protecting plant germplasm: Alternatives to patent and plant variety protection.** Jeffrey L. Ihnen and Robert J. Jondle.

11:20 **Discussion.**

11:30 **Issue and challenges in the administration of the patent law with regard to plants by the Patent and Trademark Office and of the Plant Variety Protection Act by the Plant Variety Protection Office.** Charles F. Warren.

12:10 pm **Lunch. (Capistrano A/B)**

Work Group Orientation and Goals

1:30 **Update on legislative issues pertinent to work group discussions.** Robert L. Myers.

1:45 **Ground rules, format, and organization for work group discussions.** Donald N. Duvick, Calvin O. Qualset, and Robert F Barnes.

2:00 **Work group chairs present agenda for discussion and work group sessions.**

3:15 **Break.**

3:45 **Work group sessions.**

- I. a. Room: Santa Barbara
  - b. Room: Ventura
- II. a. Room: Balboa A
  - b. Room: Balboa B
- III. Room: Balboa C

6:00 **Dinner. (On your own)**

7:30 **All delegates assemble for progress reports. (Laguna A)**

8:00 **Continuation of work group discussions.**

Thursday 2 February

8:00 am **All delegates assemble for progress reports. (El Capitan A/B)**

8:30 **Continuation of work group discussions.**

- I. a. Room: Mezzanine #1
  - b. Room: Mezzanine #2
- II. a. Room: Mezzanine #3
  - b. Room: Mezzanine #4
- III. Room: Mezzanine #5

10:15 **Break.**

10:45 **Continuation of work group discussions.**

12:00 pm **Lunch. (Capistrano A/B)**

1:30 **All delegates assemble. Work group chairs report major findings. (El Capitan A/B)**

Discussion.

3:00 **Break.**

3:30 **Work groups continue deliberations.**

6:00 **Dinner. (On your own)**

7:30 **Work groups prepare reports.**

Friday 3 February

General Assembly (El Capitan A/B)

8:30 am **Final reports from work groups and discussion.**

10:00 **Break.**

10:15 **Continuation of final reports and discussion.**

11:30 **Discussion of plans for a Public Symposium and Forum on Plant Patenting, Washington, DC, 4 May 1989.**

Closing of Workshop

1:00 pm **Steering Committee and workshop leaders edit workshop proceedings report. (Mezzanine #1)**

Support:

Science and Education, U.S. Department of Agriculture
Division of Agriculture and Natural Resources, University of California
American Society of Agronomy
Crop Science Society of America
Soil Science Society of America

Steering Committee:

Calvin O. Qualset, cochair
Donald N. Duvick, cochair
Robert F Barnes
John H. Barton
Richard L. Lower
Richard K. Perrin
Lisa J. Raines
Henry L. Shands
Howard Silverstein
Sidney B. Williams, Jr.
Paul J. Fitzgerald, ex officio

Appendix 2. List of Participants

Gary Arthur
Plant Breeder
Holden's Foundation Seeds, Inc.
R.R. 1, Box 149
Franklin, IN 46131

Robert F Barnes
Executive Vice President
American Society of Agronomy
677 S. Segoe Road
Madison, WI 53711

John H. Barton
Professor of Law
Stanford Law School
Stanford, CA 94305

Stephen A. Bent
Attorney
Foley & Lardner et al.
1800 Diagonal Road
Suite 510
Alexandria, VA 22314

Orville G. Bentley
Assistant Secretary, Science and Education
U.S. Dept. of Agriculture
Administration Bldg.
Washington, DC 20250

Fredrick A. Bliss
Professor of Pomology
Department of Pomology
University of California
Davis, CA 95616

Gardner M. Brown, Jr.
Professor and Chair
Department of Economics
University of Washington
Seattle, WA 98195

L. J. (Bees) Butler
Extension Agricultural Economist
Dept. of Agric. Economics
University of California
Davis, CA 95616

Frederick H. Buttel
Professor of Rural Sociology
Dept. of Rural Sociology
Cornell University
Ithaca, NY 14853-7801

Billy E. Caldwell
Assistant Director
North Carolina Extension Service
North Carolina State University
Raleigh, NC 27695-7602

Ronald Cantrell
Director, Maize Program
CIMMYT
Apdo Postal 6-641
Lisboa 27
06600 Mexico, DF, Mexico

James L. Carnes
President and General Manager
International Seeds, Inc.
P.O. Box 168
Halsey, OR 97348

Forrest L. Collins
Senior Patent Attorney
The Lubrizol Corporation
29400 Lakeland Blvd.
Wickliffe, OH 44092

D. C. Coston
Associate Director
South Carolina Agric. Exp. Stn.
104 Barre Hall
Clemson University
Clemson, SC 29631-0351

Dana G. Dalrymple
Research Advisor
Directorate for Food & Agric.
Bureau for Science & Technology
U.S. Agency for International Development
Department of State
Washington, DC 20523

Donald N. Duvick
Vice President for Research
Pioneer Hi-Bred International, Inc.
700 Capitol Square
400 Locust Street
Des Moines, IA 50309

Steve A. Eberhart
Director, National Seed Storage Lab.
U.S. Dept. of Agriculture
Agricultural Research Service
Fort Collins, CO 80523

Kenneth H. Evans
Commissioner, Plant Variety Protection Office
U.S. Dept. of Agriculture
Agricultural Marketing Service
National Agric. Library Bldg., Rm. 500
10301 Baltimore Blvd.
Beltsville, MD 20705

Robert E. Evenson
Professor of Economics
Economic Growth Center
Yale University
27 Hillhouse Ave.
New Haven, CT 06521-1987

Paul J. Fitzgerald
Agricultural Science Advisor
U.S. Dept. of Agriculture
Agricultural Research Service
Agronomy Dept., Poultry Sci. 202
Purdue University
West Lafayette, IN 47907

David T. Fox
Patent Examiner, Plant Biotechnology
U.S. Dept. of Commerce
Office of Patents & Trademarks
Crystal Plaza 2
Arlington, VA 22202

Kenneth J. Frey
Distinguished Professor
Dept. of Agronomy
Iowa State University
Ames, IA 50011

N. M. Frey
Director, Technology Acquisition
& Development
Pioneer Hi-Bred International, Inc.
Plant Breeding Div. POB 38
Johnston, IA 50131

I. L. (Pep) Fuller
Director, Chemical and Advanced
Technology Trade Policy
Office U.S. Trade Representative
Executive Office of the President
600 17th Street, NW
Washington, DC 20506

Burle G. Gengenbach
Professor of Agronomy
Director, Plant Molecular Genetics Institute
Dept. Agronomy & Plant Genetics
University of Minnesota
St. Paul, MN 55108

John Gerard
President
J.G. Ltd., Inc.
P.O. Box M
Plymouth, IN 46563

Rebecca Goldburg
Scientist
Environmental Defense Fund
257 Park Ave., South
New York, NY 10010

Barry Greengrass
Vice Secretary-General UPOV
34, Chemin-des-Colombettes
1211 Geneva-20
Geneva, Switzerland

Bryan Harvey
Professor and Head
Dept. of Crop Science & Plant Ecology
University of Saskatchewan
Saskatoon, Saskatchewan, Canada S7N 0W0

Robert E. Heiner
Vice President, Research & Development
Nickerson American Plant Breeder Inc.
P.O. Box 2955
Mission, KS 66201-1355

Robert W. Herdt
Director
Agricultural Sciences
The Rockefeller Foundation
1133 Avenue of the Americas
New York, NY 10036

Maureen K. Hinkle
Director, Agricultural Program
National Audubon Society
National Capitol Office
801 Pennsylvania Ave., SE
Washington, DC 20003

H. Dieter Hoinkes
Attorney
U.S. Dept. of Commerce
Box 4
U.S. Patent & Trademark Office
Washington, DC 20231

Bruce Hunter
Vice President, Research
Funks Seed International
P.O. Box 18300
Greensboro, NC 27419

Jeffrey L. Ihnen
Attorney
Robbins & Laramie Law Firm
2100 Pennsylvania Ave., NW
Washington, DC 20037

Floyd S. Ingersoll
Executive Vice President
Illinois Foundation Seeds
P.O. Box 722
Champaign, IL 61820

Clive James
Consultant, Agricultural Research
& Development
CIMMYT
Apdo Postal 6-641
Lisboa 27
06600 Mexico, DF, Mexico

Jules Janick
Professor of Horticulture
Horticulture Department
Purdue University
West Lafayette, IN 47907

John E. Kittle
Director, Patent Examining
Group 180, Biotechnology
U.S. Dept. of Commerce
Office of Patents & Trademarks
Washington, DC 20231

Roger A. Kleese
Vice President, Plant Research
& Development
Molecular Genetics, Inc.
10320 Bren Road East
Minnetonka, MN 55343

A. David Kline
Chair
Dept. of Philosophy
403 Ross Hall
Iowa State University
Ames, IA 50011

David M. Kral
Vice President, Management
and Operations
American Society of Agronomy
677 South Segoe Road
Madison, WI 53711

Warren E. Kronstad
Professor, Plant Breeding and Genetics
The Center for Gene Research
& Biotechnology
Crop Science Department
Oregon State University
Corvallis, OR 97331

William B. Lacy
Professor of Rural Sociology
Department of Sociology
University of Kentucky
Lexington, KY 40506
After 1 July 1989:
Associate Dean
College of Agriculture
Agricultural Admin. Bldg.
The Penn State University
University Park, PA 16802

Elizabeth Lassen
Director, Patents & Licensing
Calgene, Inc.
1910 5th Street
Davis, CA 95616

Victor L. Lechtenberg
Associate Director
Agricultural Research
Office of the Director
Purdue University
West Lafayette, IN 47907

William H. Lesser
Associate Professor of Marketing
Dept. of Agric. Economics
Cornell University
Ithaca, NY 14853

Richard L. Lower
Associate Dean
College of Agric. & Life Science
University of Wisconsin
Madison, WI 53706

Charles D. McLung
Patent Attorney
Chernoff, Velhuer, et al.
600 Ben Franklin Plaza
One S.E. Columbia
Portland, OR 97258

Foil W. McLaughlin
Executive Vice President
Assoc. of Off. Seed Cert. Agencies
3709 Hollsborough Street
Raleigh, NC 27607

Laura Meagher
Consultant
Center for Agric. Molecular Biol.
P.O. Box 231
Rutgers University
New Brunswick, NJ 08903

Jerry Miksche
National Program Leader, Plant Physiology
U.S. Dep. of Agriculture
Agricultural Research Service
Bldg. 005, BARC-West
Beltsville, MD 20705

Robert L. Myers
Congressional Science Fellow
U.S. House of Representatives
Committee on Agriculture
1430 Longworth Bldg.
Washington, DC 20515

Merle H. Niehaus
Dean
College of Agricultural Sciences
Colorado State University
Fort Collins, CO 80523

Phillip A. O'Berry
Principal Scientist for Biotechnology
U.S. Dept. of Agriculture
Agricultural Research Service
Rm. 101, Bessey Hall
Iowa State University
Ames, IA 50011

Kevin W. O'Connor
Study Director and Legal Analyst
U.S. Congress
Office of Technology
Washington, DC 20510-8025

Richard K. Perrin
Professor
Dept. of Economics & Business
North Carolina State University
Raleigh, NC 27695

Calvin O. Qualset
Director, Genetic Resources
Conservation Program
Div. Agric. & National Resources
University of California
1333 Research Park Drive
Davis, CA 95616

Lisa J. Raines
Director of Government Relations
Industrial Biotechnology Assoc.
1625 K Street, NW
Suite 1100
Washington, DC 20006

Thomas B. Rice
Executive Vice President
DeKalb-Pfizer Genetics
3100 Sycamore Road
DeKalb, IL 60115

Daniel Rossi
Associate Dean/Associate Director
Rutgers University
Cook College
New Brunswick, NJ 08903

J. Neil Rutger
Associate Area Director, Mid South Area
U.S. Dept. of Agriculture
Agricultural Research Service
P.O. Box 225
Stoneville, MS 39776

William T. Schapaugh
Executive Vice President
American Seed Trade Assoc., Inc.
Suite 964, Executive Bldg.
1030 15th St., NW
Washington, DC 20005

Henry L. Shands
National Program Leader for Germplasm
U.S. Dept. of Agriculture
Agricultural Research Service
Rm. 140, Bldg. 005, BARC-West
Beltsville, MD 20705

Howard Silverstein
Deputy Assistant General Counsel for Patents
U.S. Dept. of Agriculture
Office of the General Counsel
Washington, DC 20250

David Sleper
Professor of Agronomy
Department of Agronomy
University of Missouri
Columbia, MO 65211

Ann Sorensen
Assistant Director
Natural & Environmental Resources Division
American Farm Bureau Federation
255 Touhy Ave.
Park Ridge, IL 60068

Judith I. Stallmann
Assistant Professor
Dept. of Agricultural Economics
Virginia Polytechnic Inst. & State Univ.
Blacksburg, VA 24060

Gerald G. Still
Director, Plant Gene Expression Center
U.S. Dept. of Agriculture
Agricultural Research Service
800 Buchanan Street
Albany, CA 94710

John Studebaker
President & CEO Agrigenetics, Inc.
Centre Plaza North
35575 Curtis Blvd. Suite 300
East Lake, OH 44094

William H. Tallent
Assistant Administrator
U.S. Dept. of Agriculture
Agricultural Research Service
Administration Bldg.
14th & Independence Ave., SW
Washington, DC 20250

Anja van der Neut
Acting Chief, Dept. for Registration
and Plant Breeders' Rights
The Government Institute for Research
on Varieties of Cultivated Plants
P.O. Box 32
6700 AA Wageningen
The Netherlands

Charles E. Van Horn
Deputy Solicitor
U.S. Dept. of Commerce
Office of Patents & Trademarks
Box 8
Washington, DC 20231

Keith A. Walker
Vice President for Research
Plant Genetics, Inc.
1930 Fifth Avenue
Davis, CA 95616

Charles F. Warren
Supervisor Patent Examiner
Group 180, Biotechnology
U.S. Dept. of Commerce
Office of Patents & Trademarks
Washington, DC 20231

Kenneth A. Weber
Attorney
Townsend & Townsend
One Market Plaza
San Francisco, CA 94105

Ed D. Weimortz
President
Cal/West Seeds
P.O. Box 1428
Woodland, CA 95695

H. Garrison Wilkes
Professor
Department of Biology
University of Massachusetts
Boston, MA 02125-3393

Appendix 3. Work Group Organization

Participants were assigned to one of five work groups charged with evaluating specific issues associated with plant patenting. The leadership responsibilities for each work group, provided by individuals designated as the chair, facilitator, and rapporteur, are outlined as follows:

Chair—The chair provided leadership for organizing the group, conducting the group meetings, preparing the group report, and assisted in the final editing of the work group report. The chair was responsible for preparing an introductory statement on the issues and questions to be addressed by the work group.

Facilitator—The facilitator assisted and aided the chair and rapporteur in the conduct of the work group activities, including the preparation and editing of the group reports.

Rapporteur—The rapporteur took notes, recorded the pertinent findings of the group, prepared the initial group report, and assisted in editing the final group report.

The assigned topic and membership of each work group are presented below.

I. Impacts of plant patenting on research and development processes, including exchange of germplasm, property rights, international competitiveness, and public-private relationships.

A. From biotechnology viewpoint.
Chair: I.L. (Pep) Fuller
Facilitator: Laura Meagher
Rapporteur: Elizabeth Lassen

L.J. Butler
David Fox
Burle Gengenbach
Rebecca Goldburg
Bruce Hunter
W.E. Kronstad
Phillip O'Berry
Kevin O'Connor
Lisa Raines
Tom Rice
W.T. Schapaugh

B. From plant breeding viewpoint.
Chair: Roger A. Kleese
Facilitator: Richard L. Lower
Rapporteur: J. Neil Rutger

Fred Bliss
John Gerard
Bryan Harvey
J.L. Ihnen
Charles Warren
Vic Lechtenberg
Foil McLaughlin
Richard Perrin
Ann Sorenson
Keith Walker
Jerry Still

II. Impacts of plant patenting on producers and the public good.

A. Economic, social, and ethical issues.
Chair: Robert E. Evenson
Facilitator: Judith I. Stallmann
Rapporteur: William Lesser

Forrest Collins
D.C. Coston
Kenneth Evans
Robert Heiner
Floyd Ingersoll
Clive James
John Kittle
David Sleper
John Studebaker
William Tallent
Kenneth Weber

B. Plant genetic resources and environmental issues.
Chair: H. Garrison Wilkes
Facilitator: Robert Myers
Rapporteur: Gardner Brown

Gary Arthur
Billy Caldwell
Ronald Cantrell
James Carnes
Steve Eberhart
Nick Frey
Jules Janick
Merle Niehaus
Howard Silverstein
Anja van der Neut
Charles Van Horn

III. Impacts of plant patenting on international trade, genetic resource exchange, and utilization by developed countries, role of international agricultural research centers, economic effects on developing-country farmers and urban populations; also impacts of developing and developed countries' policies on germplasm exchange and utilization.

Chair: Robert Herdt
Facilitator: Fred Buttel
Rapporteur: Dana Dalrymple

Kenneth Frey
Barry Greengrass
Maureen Hinkle
Dieter Hoinkes
David Kline
William Lacy
Chuck McClung
Jerry Miksche
David Rossi
Henry Shands
Ed Weimortz

The following questions were considered by all work groups:

What are the implications, positive and negative, of Section 101 patents on plants, material characterized by plant traits, and the genetic make-up of plants?

- For patenting plants, what provision, if any, should be made for offspring research, and farmers' rights?
- For patenting material characterized by plant traits, what limits should be placed on traits or coverage?

- For patenting genes, under what conditions should they be allowed? What clarification is needed with respect to questions such as:
  - What genes are patentable—naturally occurring genes or only engineered genes?
  - Can genes transferred between distantly related species by molecular-aided techniques be considered differently from gene transfers made by conventional sexual means?
- For patents on processes, should there be any limits on the types of processes, or product of processes, that can be patented?

Work Group Reports—An edited version of each work group report is available, at cost, upon request from the CSSA Headquarters Office. Address requests to:

Executive Vice President
Crop Science Society of America
677 South Segoe Road
Madison, WI 53711

Appendix 4. Proposed Educational Program on Intellectual Property Rights for Plants

It was proposed that a significant effort be initiated for the development and implementation of an educational program on IPR for plants. An organizational structure for such a program was not developed. However, suggestions were made on potential candidates to assist in the development and implementation of such a program, as well as potential audiences for any products. These suggestions are to serve as examples only and are not intended to be complete nor in priority order. It is urged that an officer or representative from the executive office of those interested in assisting with such a program serve as a contact point and take the lead in organizing such an effort.

Suggested approaches for use in developing an educational program might include:

- Review articles in *Science*, *Nature*, *Diversity*, and other scientific and trade journals
- Preparation of bibliographic information on IPR
- Question and answer brochure on IPR
- Special conferences, fora, and workshops
- Preparation of news briefs for the communications media
- Newsletters
- Speakers bureau
- Symposia at annual meetings of scientific and professional societies and associations

Potential agencies, organizations, and groups/individuals that might be involved in the development and implementation of an educational program include the following:

Agricultural Research Institute
American Agricultural Economics Association
American Association for the Advancement of Science
American Bar Association
American Intellectual Property Law Association
American Seed Trade Association
American Society of Agronomy
Association of Official Seed Certifying Agencies
Commodity groups
Crop Science Society of America
Experiment Station Committee on Policy (ESCOP)—state agricultural experiment stations
Extension Committee on Policy (ECOP)—federal and state extension service
Farm organizations (American Farm Bureau Federation, National Farmers Organization, National Farmers Union, National Grange)
Industrial Biotechnology Association
National Council of Commercial Plant Breeders

National Research Council, Board on Agriculture
National Plant Germplasm System
Office of Technology Assessment
University Departments of Agronomy, Botany, Biological Sciences, Crop Science, Agricultural Economics, Horticulture, Microbiology, Pomology, Sociology, and Soil Science
UPOV—International Union for the Protection of New Varieties of Plants
U.S. Agency for International Development
USDA—Agricultural Marketing Service
USDA—Agricultural Research Service
USDA—Plant Variety Protection Office
USDC—Patent and Trademarks Office
World Bank

Potential audiences for receiving information on IPR for plants include the following:

Congress (congressional hearings, circulation of reports, visits, provide list of competent and available speakers to address policy issues)
County extension agents (contact through federal, state, and county offices)
Citizen groups (articles in *Diversity*; workshops, symposia and fora on key topics; outlet through media of newspaper, radio, and television)
Developing countries (international conferences; communication through IARC, such as CIMMYT, USAID, and organizations of the United Nations, such as FAO)
Farmers (farm publications, communications with farm organizations)
Heads of appropriate departments within universities
Plant breeders (private and public sector)
Research scientists (private and public sector. Contact may be through the appropriate scientific and educational societies.)
Seed industry (communicate with directors of research and development)
State agricultural experiment stations (communicate with individual SAES directors, and Experiment Station Committee on Policy—ESCOP)